AF789554

Adaptations of Desert Organisms

Edited by J. L. Cloudsley-Thompson

Springer

Berlin
Heidelberg
New York
Barcelona
Budapest
Hong Kong
London
Milan
Paris
Santa Clara
Singapure
Tokyo

Volumes already published

Ecophysiology of the Camelidae and Desert Ruminants
By R. T. Wilson (1989)

Ecophysiology of Desert Arthropods and Reptiles
By J. L. Cloudsley-Thompson (1991)

Plant Nutrients in Desert Environments
By A. Day and K. Ludeke (1993)

Seed Germination in Desert Plants
By Y. Gutterman (1993)

Behavioural Adaptations of Desert Animals
By G. Costa (1995)

Invertebrates in Hot and Cold Arid Environments
By L. Sømme (1995)

Energetics of Desert Invertebrates
By H. Heatwole (1995)

Ecophysiology of Desert Birds
By G.L. Maclean (1996)

In preparation

Plants of Desert Dunes
By A. Danin (1996)

Gordon Lindsay Maclean

Ecophysiology of Desert Birds

With 69 Figures

Springer

Professor Dr. Gordon Lindsay Maclean
University of Natal
Department of Zoology and Entomology
Private Bag X01
3209 Scottsville
South Africa

Cover illustration: Photograph by J.L. Cloudsley-Thompson

ISBN-13:978-3-642-64639-3 Springer-Verlag Berlin Heidelberg New York

Library of Congress Cataloging-in-Publication Data
Maclean, Gordon L.
The ecophysiology of desert birds / Gordon Lindsay Maclean.
p. cm. -- (Adaptation of desert organisms)
Includes bibliographical references (p.) and Index.
ISBN-13:978-3-642-64639-3 e-ISBN-13:978-3-642-60981-7
DOI: 10.1007/978-3-642-60981-7

1. Birds--Ecophysiology. 2. Desert fauna--Ecophysiology.
3. Birds--Adaptation. 4. Desert fauna--Adaptation. I. Title.
II. Series.
QL698.95.M33 1995
598.29'15'4--dc20 95-34589

This work is subject to copyright. All rights are reserved, whether the whole or part of the material is concerned, specifically of translation, reprinting reuse of illustrations, recitation, broadcasting, reproduction on microfilm or in any other way, and storage in data banks. Duplication of this publication or parts thereof is permitted only under the provisions of the German Copyright Law of September 9, 1965, in its current version, and permission for use must always be obtained from Springer-Verlag. Violations are liable for prosecution under the German Copyright Law.

© Springer-Verlag Berlin Heidelberg 1996
Softcover reprint of the hardcover 1st edition 1996

The use of general descriptive names, registered names, trademarks, etc. in this publication does not imply, even in the absence of a specific statement, that such names are exempt from the relevant protective laws and regulations and therefore free for general use.

Typesetting: K. Darms, Göttingen
SPIN 10033063 31/3137-5 4 3 2 1 0 - Printed on acid-free paper

Preface

In the Preface to his volume on *Ecophysiology of the Camelidae and Desert Ruminants* in this series, Trevor Wilson mentions his hesitation at the invitation from John Cloudsley-Thompson to write the book, because he had been out of direct touch with desert biology for some years. My response to John's invitation to write *Ecophysiology of Desert Birds* was much the same, and for the same reasons. However, with his encouragement and, taking account of the fact that a colleague of mine who is much better suited to do the job had turned it down because of pressure of work, I accepted the challenge and have enjoyed it immensely. It has been my privilege to spend much time in the deserts of North and South America, southern Africa and Australia, some of it with my wife, Cherie, whose editing of the entire text has improved the work immeasurably. Indeed, I believe it would have been proper for her to accept coauthorship, but she graciously declined the offer! I owe her a debt of gratitude which I hope will be repaid in part by her seeing the text finally between covers.

The chapters on water regulation (Chap. 5) and thermoregulation (Chap. 6) have benefited from the refereeing of my colleague, Dr. Barry Lovegrove, of the Department of Zoology and Entomology, University of Natal, Pietermaritzburg. He is an authority on desert biology and the author of the magnificent book, *The Living Deserts of Southern Africa*; as a physiologist of repute, he has greatly improved the presentation in these chapters. In the longer term, I have derived much pleasure and profit from associations over the years with Tom Cade, George Bartholomew, Tom Howell, Fred White, John Wiens, Ed Wyndham, Allen Keast, Llew Greenwald, Ernie Willoughby, David Thomas and the late Klaus Immelmann: this is an appropriate opportunity to thank these colleagues for their encouragement, interaction and publications. I hope that they, too, may derive pleasure from this book and my free use of their work.

My thanks also go to Américo Bonkewitzz for the fine computer-generated diagrams. For the use of photographs I am indebted to Mike Lawes (Negev Desert), Olaf Wirminghaus (Roadrunner), G.K. Lestrange (White-backed Mousebird and Red-billed Firefinch) and Graeme Chapman (Black-faced Woodswallow). The rest of the photographs are my own. My final word of gratitude goes to the Department of Zoology and Entomology, University of Natal, Pietermaritzburg, for facilities and support; in particular I wish to thank Billy Boodhoo for his skill in making black-and-white prints from my negatives and colour slides.

Pietermaritzburg, South Africa,
October 1995

GORDON LINDSAY MACLEAN

Contents

Abbreviations and Symbols used in this Book

ADH	= antidiuretic hormone
AMR	= active metabolic rate
ANGII	= angiotensin II
AVT	= arginine vasotocin
BMR	= basal metabolic rate
CEWL	= cutaneous evaporative water loss
CWL	= cutaneous water loss
EHL	= evaporative heat loss
EWL	= evaporative water loss
FMR	= field metabolic rate
GFR	= glomerular filtration rate
I_{dw}	= dietary water index
M	= molar (molarity; 100% sea water is 0.5-0.6 M NaCl; a solution of 37.5% sea water has a P_{osm} of 354.6 mOsmol/l)
mEq	= milliequivalents
mEq/l	= milliequivalents per litre
mmol/l	= millimoles per litre
mOsmol	= milliosmoles
mOsmol/l	= milliosmoles per litre
mW/g	= milliwatts per gram (metabolic rate)
P_{osm}	= serum or plasma osmotic pressure (normally ±370-390 mOsmol/l for most birds)
REWL	= respiratory evaporative water loss
RH	= relative humidity
RMR	= resting metabolic rate
SMR	= standard metabolic rate
T_A	= ambient temperature
T_B	= body temperature
T_{LC}	= lower critical temperature
TNZ	= thermoneutral zone
T_{UC}	= upper critical temperature.
U_{osm}	= urine osmotic pressure
WEI	= water-economy index
WIR	= water-influx rate

1 Introduction

For the purposes of this book, any land area outside of the polar regions, which receives 250 mm of rain or less per year will be called a desert. It has been customary to divide deserts into semi-arid (somewhere between 125 and 250 mm annual rainfall) and arid (less than 125 mm annual rainfall) (Maclean 1974c), and these subdivisions may be loosely useful, though some so-called hot deserts, such as that of Madagascar, may have an annual precipitation of as much as 600 mm (Evenari 1985). Deserts cover about 25-30% of the land surface of the planet (Polis 1991); they have low and erratic rainfall, and loss of water by evaporation usually exceeds that gained by precipitation. Primary productivity is low: the lowest standing biomass ever recorded was in the Namib Desert, Namibia, after a drought (Seely and Louw 1980). However, desert ecosystems are not as simple as may first appear; they are relatively complex and biologically rich, though diversity decreases with increasing aridity (Polis 1991).

Hot deserts are created mostly in the subtropical belt of high pressure roughly between latitudes 20° and 40° N and S, but other factors in their development or exacerbation include rain shadows caused by mountain barriers, cold air from adjacent cold ocean currents, continentality and possibly high albedo (Evenari 1985).

My analysis of the literature on desert birds as presented here shows that the heyday of arid-zone ornithology seems almost to have passed, though it showed a flicker of revival in the early 1990s. My survey of the literature, though it may not be exhaustive, shows that there is in fact a surprising paucity of literature on the ecology of arid-zone birds, despite the fact that birds lend themselves admirably to work in both field and laboratory: there were only four significant publications in the 1940s and only seven in the 1950s, but there was a sudden and dramatic increase in the following three decades (58 publications in the 1960s, 82 in the 1970s and 54 in the 1980s). Certain years were especially productive, notably 1966-1973, 1976 and 1977 (Fig. 1.1). Of the literature surveyed here, so far the 1990s have produced only 23 relevant publications, though the decade is already 6 years old at the time of writing.

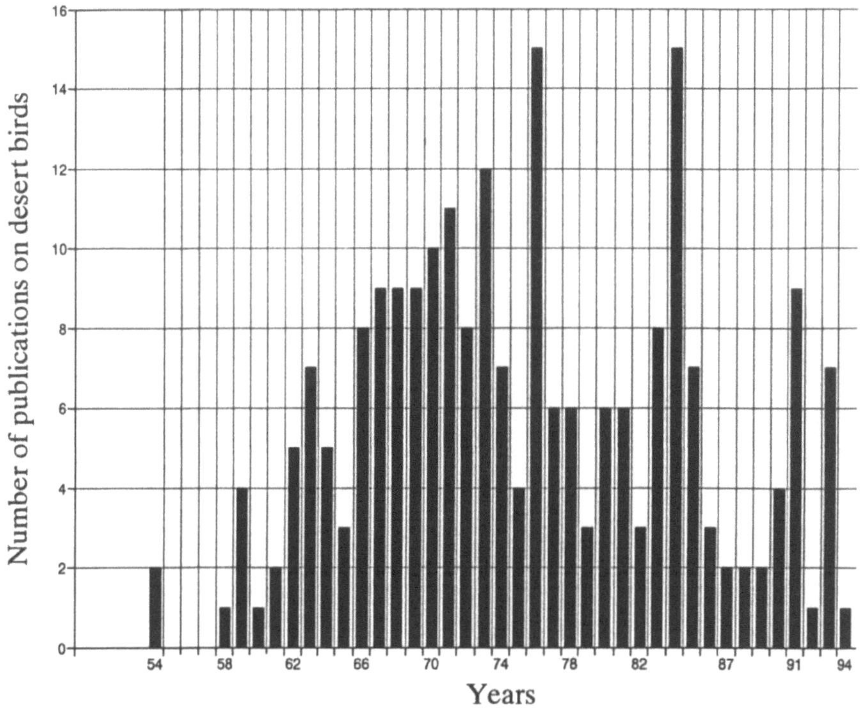

Fig. 1.1. Number of publications on arid-zone ornithology per year since 1954, as reviewed in this book

My own involvement with deserts began in the late 1950s when I worked on a diamond mine in the Namib Desert (Fig. 1.2), but it was not until 1964 that I was able to pursue serious research on arid-zone birds, this time in the Kalahari (Fig. 1.3). Since then I have also had the privilege of observing birds in the deserts of North and South America (Figs. 1.4 and 1.5) and of Australia (Fig. 1.6).

Several general reviews of arid-zone ornithology have already appeared (e.g. Schmidt-Nielsen 1964, pp 204-224; Dawson and Bartholomew 1968; Serventy 1971; Skadhauge 1972; Davies 1976; Dawson 1976, 1984; Thomas 1984b), as have some of a more regional nature, such as Casselton (1984) on the Sahara, Maclean (1974c, 1976c, 1984c) on southern Africa and South America, Immelmann (1963a) and Dawson (1981) on Australia, and Mares et al. (1977) on the Americas. Certain papers have dealt with special groups or species of birds, notably the sandgrouse (Maclean 1976d, 1985; Thomas 1984a, c), the Sociable Weaver *Philetairus socius* (Maclean 1973b-f) and waders (Maclean 1984a). The present work is an attempt to draw the information from these works and from the scattered journal literature into a coherent whole, from which the reader may gain both a global and a detailed picture of the ways in which birds are adapted to the arid environment.

Fig.1.2. Sand sea, Namib Desert, Namibia

Fig. 1.3. Kalahari sandveld, Northern Cape, South Africa

Fig. 1.4. Sonoran Desert near Tucson, Arizona, USA

Fig. 1.5. Central Atacama Desert, Chile, South America

Fig.1.6. Central Australian Desert, northwestern New South Wales

The term "ecophysiology" encompasses such a wide range of biological subdisciplines that almost anything adaptive has to be captured by the ecophysiological net, whether it be behavioural or structural, strictly ecological (relating the organism to its environment), or strictly physiological (investigating the organism's internal physicochemistry). All these aspects of an organism's being involve some adaptiveness, at least a part of which will be directly or indirectly related to the desert habitat and the stresses it puts on the organism.

Apart from some work on North American birds by Kendeigh (1939, 1944), Seibert (1949), Vorhies (1928, 1945) and Sumner (1935), there are no significant quantitative or experimental studies on the water economy of wild land birds earlier than 1950 (Bartholomew and Cade 1963). Bartholomew and Cade's (1963) review is not confined to desert birds, but they give a comprehensive review of ecological patterns of water economy in desert birds, followed by an outline of research problems current at the time; many are still current. The ecological patterns of desert birds fall into three categories: (1) obligate drinkers such as the House Finch *Carpodacus mexicanus* and House Sparrow *Passer domesticus* which never venture far from water; (2) birds such as doves which are dependent on surface water for drinking, but capable of drinking only occasionally because of good powers of flight and a large alimentary capacity; and (3) largely carnivorous and insectivorous birds which obtain all their water needs from their succulent food and do not therefore need to drink. This review

(Bartholomew and Cade 1963) has shown that not much was known about the water economy of land birds at the time. Some of the research problems posed include:

1. Determination of rates of dehydration and water consumption;
2. Minimum daily water requirements for a bird on a dry diet under conditions of known temperature and humidity;
3. Ability of birds to discriminate between naturally occurring waters on the basis of taste;
4. Effects of salinity, temperature and humidity on birds;
5. Comparison of osmotic pressures and electrolyte concentrations of birds in different taxa;
6. Structure and function of avian kidneys;
7. Adaptations of the gut and cloaca to water conservation;
8. Determination of total body water using ethanol dilution under different conditions of hydration;
9. Rates of water turnover using tritiated water;
10. Relationships between metabolic rate, ambient temperature (T_A) and respiratory water loss;
11. Drinking needs and frequencies of different birds;
12. Temperature, humidity and air movements in closed nests and nest holes;
13. Use by birds of burrows and other shelters in the heat of the day;
14. Possible seasonal shifts in diet, such as to more succulent foods in the dry season, and so on.

The following preadaptive characteristics of birds have operated to refine the water economy of birds (Bartholomew 1972):

1. Energy Metabolism. Birds have high rates of energy metabolism. The rate for a small passerine may go up to 10-25 times the resting metabolic rate in flight, sometimes for hours at a time.
2. Body Temperature. The characteristically high body temperatures (T_B) of birds allow them to lose heat to the environment by convection at most ambient temperatures, removing a primary dependence on evaporative cooling.
3. Evaporative Water Loss. Birds have high rates of pulmocutaneous evaporative water loss, a consequence of their high metabolic rates and high T_B.
4. Excretion. Nitrogenous wastes are excreted in the form of uric acid, and many species excrete excess electrolytes via the nasal (salt) gland.
5. Diurnality. Unlike most mammals, birds are largely diurnal, which can pose problems of water balance and thermoregulation.
6. Mobility. Birds are among the most mobile of terrestrial animals, allowing them to escape from environmental problems in ways not open to walking animals; however, this may also exacerbate problems of water balance, especially during migration or longer nomadic movements.

A source of water conservation in at least some birds, to which little attention has been paid, is the countercurrent effect produced in the nasal passages by inhaled and exhaled air. Exhaled air is saturated with water vapour; the higher its temperature, the more water it carries and the greater the degree of evaporative water loss (EWL). At T_A below T_B inhaled air cools the respiratory tract, resulting in subsequent cooling of exhaled air and some condensation of its water vapour. The lower the temperature of the respiratory tract, the greater the amount of water vapour recovered. In the North American Cactus Wren *Campylorhynchus brunneicapillus* at T_A below 35 °C the temperature of exhaled air is always less than that of inhaled air, so that countercurrent cooling must be important in water conservation. Since this mechanism seems to operate in all birds in all kinds of habitats (including, for example, the Ostrich *Struthio camelus*; Withers et al. 1981), it is a useful preadaptation to life in the arid zone, and not a new feature of desert birds in particular.

Since few desert birds are nocturnal, most are exposed to the full stress of the desert environment (Wiens 1991). Ground-nesters are under even more stress than other kinds. As endotherms, birds have high energy demands, many of which are met by feeding on seeds, but the cost of such a path is twofold: (1) seeds are a patchy resource in both time and space since they depend on rainfall for their production and this may be localized; and (2) air-dried seeds contain little water (usually less than 15%). Another preadaptation of birds to the arid environment is their high mobility; others include a relatively high T_B (around 40 °C) and an ability to tolerate hyperthermia. Also preadaptive is the excretion of uric acid as the nitrogenous waste product, which allows birds to use less than 10% of the amount of water required by mammals to excrete an equivalent amount of nitrogen.

Generally speaking, animals living in a semi-arid region may be more rain-dependent for some of their activities, notably breeding, than are those inhabiting really arid regions where the amount of rain is so small and the timing of rain so erratic as to be an unreliable environmental cue for anything, except for highly nomadic species whose movements are partly governed by rain. The high degree of mobility enjoyed by birds preadapts them admirably to a nomadic existence, and there are many nomadic species, but a significant number of species of birds in such extremely arid deserts as the Namib and Atacama are resident, if not actually sedentary. Further examples exist in Australia: few Australian birds are true migrants (about 8%), but at least 26% are nomadic (Keast 1959). Even some true desert birds are nevertheless sedentary (*Amytornis* spp., *Oreoica gutturalis*, *Malurus leucopterus*, *M. callainus*, and *Artamus melanops*). The number of species of birds in any given area may vary from 10 in a drought year to over 90 in a wet year.

A general overview of the physiological adaptations of birds to aridity will examine these in relation to temperature, endothermy and thermoregulation, excretory functions, and osmoregulation. Two chapters are devoted solely to water in the arid environment: its sources, chemistry and physiological processing. Thermoregulation is then also investigated in greater detail. An especially important aspect of desert-adapted birds is their ability to reproduce under severe conditions; breeding biology is therefore examined in depth. The factors initiating breeding behaviour and influencing gonadal cycles, both intrinsic and extrinsic, are analyzed as far as present evidence allows. The ecology of nest sites, nest temperatures and incubation strategies then come under scrutiny. An examination of post-hatching parental care concludes the ecophysiology of breeding. Moulting is a particularly stressful process in the desert biome, and has to be modified so as to call for as little energy input on the part of the bird as the habitat will allow, especially during breeding.

The particular problems of life in hot deserts may be summarized as low biological productivity, scarcity of free water, and regularly high environmental heat loads (Thomas and Maclean 1981). Conservation in the deserts of the world is also a major problem where settlement by humankind has occurred: Dean et al. (1991) have shown, for example, that heavy grazing in the restricted habitat of the Red Lark *Mirafra burra* in the northwestern Cape of South Africa has made large areas unsuitable for its continued existence. Protection of the remaining habitat is a priority for the conservation of the species.

1.1 Preadaptation of Birds to the Arid Zone

From what has been cited above from Bartholomew (1972) and other workers, much of what seems to be adaptive in birds to the desert environment is in fact intrinsic to the avian condition. The study by Chappell et al. (1984) on the California Gull *Larus californicus* breeding on barren islets in Mono Lake, California, where it is subjected to intense solar radiation, shows this clearly. The gulls use behavioural responses rather than special physiological adaptations to cope with high T_A. Chicks have good thermoregulatory ability quite soon after hatching and can withstand T_A from 6.0-42.5 °C without signs of distress, even when as small as 70 g body mass. Above T_A 43 °C the chicks become frantic, but are usually shaded by the parents under natural conditions. The physiological capabilities of these gulls are no different from those of other gulls (e.g. Heermann's Gull *Larus heermanni*; Bartholomew and Dawson 1979) and other marine birds tested (e.g. Laysan Albatross *Diomedea immutabilis*, Black-footed Albatross *D. nigripes*, Howell and Bartholomew 1961; Red-tailed Tropicbird *Phaethon rubricauda* and Red-footed Booby *Sula sula*, Howell and Bartholomew 1962) and are a further indication of the preadaptiveness of birds to hot, dry climates. At Mono Lake, the T_A can reach 50-60 °C in unshaded areas on sunny days and, even when shaded by the parents, chicks would pant for long periods, sometimes up to 9 h continuously.

Birds are unsurpassed as subjects for experimental and observational research in the field and in the laboratory. This book is evidence of their remarkable capacity to survive in the desert.

2 Desert Avifaunas

Generally, levels of speciation and endemism of birds are low in deserts and the number of bird species largely confined to deserts is small (Wiens 1991). Even so, certain taxa contribute disproportionately to desert avifaunas, e.g. Burhinidae, Pteroclidae, Thinocoridae, Alaudidae, Ephthianuridae (four out of five species), some Glareolidae and Charadriidae, certain genera of Furnariidae, Tyrannidae, Mimidae and Turdidae (Maclean 1976c; Tables 2.1 and 2.2). Convergent evolution of birds of different taxa in the different deserts of the world is a phenomenon to which attention has been drawn from time to time (e.g. Maclean 1976c; Drozdov 1984); this is not surprising in view of the similarly harsh environmental selection pressures to which arid-zone birds are subjected. Such convergence is physiological, morphological and behavioural.

Table 2.1. Predominantly arid-zone avian taxa in the Afrotropical and Neotropical regions. (Maclean 1976c)

Taxon	Distribution	% of species Arid	% of species Non-arid
Burhinidae	Neotropical	100	0
Cursoriinae	Palaeotropical	52	48
Thinocoridae	Neotropical	75	25
Pteroclidae	Africa and Asia	81	19
Furnariidae (*Geositta, Cinclodes, Upucerthia, Eremobius, Chilia*)	Neotropical	60	40
Furnariidae (*Asthenes*)	Neotropical	55	45
Tyrannidae (*Muscisaxicola*)	Neotropical	64	36
Alaudidae	Africa and Asia	67	33
Turdidae (*Oenanthe, Cercomela*)	Africa and Asia	72	28
Mimidae	Neotropical	51	49
Mean % all taxa		68	32

Table 2.2. Avian taxa in the Afrotropical and Neotropical regions, that contribute substantially to arid-zone avifaunas. (Maclean 1976c)

Taxon	Distribution	% of species	
		Arid	Non-arid
Phasianidae	Central America	33	67
Otididae	Africa and Asia	40	60
Troglodytidae	Central America	33	67
Fringillidae	Neotropical	25	75
Mean % all taxa		33	67

Maclean (1970c) derived a simple formula for calculating an abundance index (A) for any bird species:

$$A = \frac{\sum M}{m}$$

where ΣM= the sum of the total of the daily field counts for a given month and m = the number of months involved in the analysis. The abundance indices for birds in the southern region of the Kalahari Gemsbok National Park were allocated categories of abundance as shown in Table 2.3 for different values of A. An analysis by number of species gives some idea of the relative frequency of seedeaters in the avian population of four arid zones (Table 2.4), but an analysis by number of individuals gives an even more impressive picture of the abundance of seedeaters, using the above formula (Table 2.5).

A monthly index (Z) for any category of abundance (A) can be calculated from the formula:

$$Z_A = \frac{\sum M_A}{S_A \; F}$$

where A = the relevant category of abundance (calculated as shown above), S = the number of species per category of occurrence (resident, nomadic, occasional, breeding migrant, non-breeding migrant) and F = the total number of field trips per month. Values of Z can be calculated for any category of abundance or for a single species or for the whole avifauna.

Table 2.3. Abundance indices and categories of abundance as calculated by Maclean (1970c) and used in Table 2.5

Category of abundance	Value of A
Abundant	>100
Very common	50-100
Common	10-50
Fairly common	4- 10
Uncommon	2- 4
Rare	1- 2
Very rare	<1

Table 2.4. An analysis of the avifaunas of four world desert regions, showing proportions of passerine species and seed-eating species

Region	No. of species observed	Passerines Number	Passerines %	Seed eaters Number	Seed eaters %	Reference
Kalahari (South Africa)	74[a]	37	50	17	23	Maclean (1970c)
Kahalari (South Africa)	52	26	50	9	17	Pianka and Huey (1971)
Sahel (Senegal)	25[a]	13	52	13	52	Morel and Morel (1973)
Western Australia	70	41	59	12	17	Pianka and Pianka (1970)
Ecuador	34	23	68	9	26	Marchant (1958)
Means			56		27	

[a] Residents and nomadic breeders only.

Table 2.5. Comparison of abundance of passerines and non-passerines, and of seed-eating and non-seed-eating birds in the southwestern Kalahari (resident and breeding nomadic species only). (Maclean 1970c)

Group	Abundance index[a]
All species	29.19
Passerines	40.79
Non-passerines	17.60
Seedeaters[b]	103.67
Non-seedeaters	13.37

[a] See Table 2.3 and text for explanation.
[b] Excluding sandgrouse, Cape Turtle Dove, Sociable Weaver and House Sparrow.

This type of analysis allows for trends to be quantified. Maclean's (1970c) study showed that, among the resident species in the Kalahari, only the Namaqua Sandgrouse *Pterocles namaqua*, Burchell's Sandgrouse *P. burchelli*, Cape Turtle Dove *Streptopelia capicola*, Sociable Weaver *Philetairus socius* and Cape Sparrow *Passer melanurus* could be classified as "abundant", all seedeaters to a greater or lesser extent. Only three species were "very common": Scaly-feathered Finch *Sporopipes squamifrons*, Yellow Canary *Serinus flaviventris* and Spike-heeled Lark *Chersomanes albofasciata* (only the lark is not primarily a seedeater). The remaining categories of resident species consisted mainly of insectivorous species and raptors, some of which were at the edge of their distributional ranges and therefore mostly "uncommon" to "very rare".

Ten species were nomadic breeding birds, all but two of which were "common" to "abundant", indicating that they occurred in very large numbers when they were present. Four of these ten species were larks, including two of the three species in the "abundant" category (Grey-backed Finchlark *Eremopterix verticalis* and Black-eared Finchlark *E. australis*). The third "abundant" species was the Namaqua Dove *Oena capensis.* Nomadism is an important feature of the behavioural ecology of desert birds in all arid zones, but especially those of the southern continents, as has also been shown, for example, in Australia by Davies (1984).

All but 3 of the 24 migrant species in the Kalahari study area, whether breeders or non-breeders, were "uncommon" to "very rare", which is not surprising in view of the unpredictable nature of deserts generally.

The species richness of birds in deserts is low, decreasing rapidly with decreasing rainfall in most deserts, but not in the North American deserts where the driest regions have great diversity of topography and landscape (Wiens 1991). Ground-feeding species predominate in the Kalahari and Sonoran Deserts, while arboreal species predominate in the deserts of Western Australia and the Argentine Monte, probably as a result of a greater abundance of trees in the latter two types of arid zone.

Of the 21 most arid-adapted resident (= sedentary?) species of birds on the Santa Elena Peninsula, Ecuador, 67% were passerine, of which 36% were suboscine (Marchant 1958). A further category of 13 species of presumed residents, also therefore likely to be arid-adapted, consisted of 69% passerines, of which 33% were suboscine. Representation of suboscines in an arid-zone avifauna is unique in the deserts of the world, since all the Old World suboscines are forest or woodland species.

In the vicinity of Fété-Olé in the Sahelian savanna of Senegal only 15 species appeared to be resident (or sedentary?) and then only in small numbers as a rule: Pteroclidae (1 sp.); Columbidae (4 spp.); Coraciidae (1 sp.); Bucerotidae (2 spp.); Lybiidae (1 sp.); Alaudidae (1 sp.); Sturnidae (2 spp.); Sylviidae (1 sp.); Ploceidae (1 sp.); and Estrildidae (1 sp.) (Morel and Morel 1973). Of these, seven are granivorous or frugivorous, six are omnivorous and two solely insectivorous. The low numbers were probably the result of a paucity of microhabitats and of a low food supply. Numbers were greater in the wet season, the extent of rain in which affected the density of birds in the following dry season. Afrotropical birds are more markedly influenced by seasonal changes than are Palaearctic species. In addition to these, only another 10 species bred in the area: Strigidae (1 sp.); Columbidae (1 sp.); Coliidae (1 sp.); Ploceidae (2 spp.); Viduidae (1 sp.); Estrildidae (1 sp.); Sylviidae (2 spp.); Nectariniidae (1 sp.).

Species diversity of birds in arid habitats dominated by Creosotebush *Larrea divaricata* in southern Arizona (Fig. 2.1), United States of America, increases with species richness and with an increasing evenness of distribution of individuals among species (Tomoff 1974), as in any other habitat. Most birds nest in cacti or other spinescent plants, all within or adjacent to tree-lined watercourses (Fig. 2.2), so that the diversity of plant structure is very important to the occurrence and breeding of desert birds. Greater physiognomic diversity of plants is reflected in a greater number of avian foraging patterns. The assumption of randomness in the selection of nest sites by birds in forest or woodland habitats is not supported by a study of birds in desert habitats, since the birds are highly responsive to particular life forms, selecting nest sites in plant species which have a particularly favourable

Fig. 2.1. Creosotebush habitat, New Mexico, USA

Fig. 2.2. Tree-lined watercourse in Arizona desert, USA

structure (e.g. cacti). This is a wide departure from random nest-site selection. The composition of plant species in desert scrub is therefore highly significant in regulating communities of breeding birds. "One major difference between deciduous forest and desert scrub is the occurrence of specific structural elements in the vegetation present in trace amounts but critical to birds for nest sites" (Tomoff 1974).

In the Kalahari Desert of South Africa only the plant-height diversity appears to be significantly correlated with bird-species density (Pianka and Huey 1971). These workers allocated the avifauna of the Kalahari to the following four foraging and food categories: (1) ground-foraging herbivores; (2) ground-foraging carnivores; (3) arboreal herbivores; and (4) arboreal carnivores. Some bird species could be assigned to two or four of these categories; an analysis of the avifauna divided in this way showed that, as overall bird-species density increased, arboreal species were added faster than ground species. An earlier study by Pianka and Pianka (1970) allowed for a comparison to be made between Kalahari and Australian bird censuses in arid regions; they found that:

1. There are proportionately more species of ground carnivores in the Kalahari than in Australia.
2. The percentages of ground herbivores are similar in both arid regions.
3. There are proportionately more species of arboreal carnivores (and arboreal species overall) in Australia than in the Kalahari.

Two conclusions to be reached here are that trees contribute disproportionately greatly to habitat diversity, and that the Australian arid regions are considerably better wooded than the southern African counterparts. The importance of trees to certain desert birds has been demonstrated by Sopyiev (1984) in the Kara-Kum Desert of Russia, where the spread of the Sand Acacia *Ammodendron conollyi* under the influence of anthropogenic changes to the habitat has assisted the spread of the Desert Sparrow *Passer simplex*. Similarly, the presence of trees is essential to the occurrence of the Sociable Weaver *Philetairus socius* in southern Africa: the density of the nest masses of these birds is a function of the density of trees, though not directly proportional to that parameter, since territorial behaviour ensures that adjacent colonies are seldom less than about 1 km apart (Maclean 1973b).

Sandgrouse are predominantly an arid-zone family of birds (Table 2.1). Their adaptation to this environment has been shown repeatedly in different studies throughout the range of the family. For example, Aldrich (1943) observed that the Chestnut-bellied Sandgrouse *Pterocles exustus* in Kaira District, India, was much more abundant in drought years than in years of good rainfall. It appeared to breed there only in the drought years, indicating an adaptation to arid conditions.

The Chestnut-bellied Sandgrouse is most abundant in the Thar and Sind Deserts of India and Pakistan, thousands of birds coming to drink at waterholes in summer when water is restricted (Christensen et al. 1964). The

Table 2.6. Mean maximum and minimum temperature ranges (°C) in the Thar and Sind Deserts of India and Pakistan. (Christensen et al. 1964)

Season	Thar Desert		Sind Desert	
	Mean max.	Mean min.	Mean max.	Mean min.
Winter (December-February)	24.4-27.2	9.4-11.7	24.4-27.2	10.6-12.2
Spring (March-May)	32.2-40.6	16.1-26.1	33.3-41.7	17.8-25.6
Summer (June-August)	33.3-39.4	25.0-27.8	35.6-40.0	26.1-27.8
Autumn (September-November)	30.6-35.0	12.8-23.9	31.7-36.7	15.0-24.4

region experiences a rainfall grading from 125 mm/year in the drier western parts to 350 mm/year in the higher-rainfall region. Where rainfall reaches 675 mm/year the flocks are smaller and more scattered. Mean temperature regimes in these two deserts are shown in Table 2.6. Air temperatures over the range of the sandgrouse can reach extreme highs of 49 °C, but minima seldom go as low as 0 °C. The main nesting month is March which is early spring, and dry.

By contrast, certain birds which live at the edge of the arid zone may have much less tolerance of the extreme conditions which a "true" desert presents. For example, Gambel's Quail *Callipepla gambelii* inhabits arid regions in the American southwest, where the mean rainfall is at least 185 mm/year (Gullion 1960). Rainfall of 124 mm/year is insufficient to produce the vegetation necessary to sustain Gambel's Quail in the deserts of Nevada, but a mean rainfall of 186 mm/year, only 480 km away in Nevada, does provide adequate vegetation. However, populations of quails increased only after falls of 262 and 298 mm; a fall of only 246 mm did not suffice to allow populations to increase. Populations are light where the mean rainfall is 254 mm/year or less.

2.1 Coloration

One of the most obvious features of many desert birds is their apparently good camouflage, deriving from the matching coloration of their plumage with that of the colour of the soils on which they live. The notion put forward by a number of observers in the past, that desert coloration is physiologically adaptive rather than cryptic, is shown to be unlikely in view of the fact that localized desert areas with differently coloured soils have matching populations of animals confined to them both spatially and behaviourally (acridid grasshoppers, geckos of the genus *Rhoptropus*, mice and larks, to name a few examples), in spite of the fact that identical climatic conditions exist in all these areas (Niethammer 1959). The Santa Elena Peninsula of Ecuador is a rather flat to undulating landscape of a generally dull brownish grey colour which is reflected in the plumages of many resident species of birds, such as the Short-tailed Groundtyrant *Muscigralla brevicauda* and the Parrot-billed Seedeater *Sporophila peruviana* (Marchant 1958). The conclusion is that selection by predators is the operative force (Niethammer 1959).

Objections to this conclusion have come from other observers who claim that the density of predators in desert is too low to be a significant factor. Niethammer (1959) is nevertheless convinced that predators are the selective force, even though he did not himself ever see a lark taken by a predator. However, other workers have seen Lanner Falcons *Falco biarmicus* and Peregrine Falcons *F. peregrinus* hunt sandgrouse (*Pterocles* species) and Sand Partridges *Ammoperdix heyi* in the Sahara. Stomach contents and regurgitated pellets of Lanners in the Sahara have contained remains of Short-toed Lark *Calandrella brachydactyla* and Cream-coloured Coursers *Cursorius cursor*. *Falco* species in North America are also known to catch small passerines (larks and meadowlarks). The Lanner Falcon in the Sahara indeed regulates its breeding to the passage of small migratory birds northward in the northern spring (April), but it is not known to what extent Lanners prey on resident desert birds at other times. It may be significant that, unlike the case in many other resident bird species, both predator and prey, there is no subspecies of Lanner confined to the Sahara, which points to a highly mobile population or to one which has only recently colonized the Sahara. The latter suggestion is unlikely because the Sahara is still expanding and presents the Lanner with no better habitat today than it did in earlier times.

Sedentary prey species of birds must suffer greater predation from raptors when the Palaearctic migrants are absent, but this may be compensated for by the fact that they usually breed in summer and by the additional recruitment to the population in the form of young. However, information on the productivity of larks in the Sahara is negligible. A Lanner is likely to eat 1460 birds the size of a lark or 300 birds the size of a sandgrouse each year. Because of the lower density of resident desert birds than migratory Palaearctics, the desert birds must be under far greater selection pressure in between the passages of migrants, than the pressure on the migrants themselves. However, the population of Lanners also drops in the summer between the migratory seasons.

Niethammer (1959) concludes unavoidably that predator pressure is the main factor operating to select for desert coloration, whether pale or dark, depending on the substrate. As just one example, the Chestnut-bellied Sandgrouse in the Thar and Sind Deserts of India and Pakistan does not seek cover for concealment, but relies on its protective coloration which "blends in well with desert sand and rocks" (Christensen et al. 1964). Furthermore, Maclean (1970a) demonstrated that the eight species of larks breeding in the Kalahari sandveld (Fawn-coloured *Mirafra africanoides*, Sabota *M. sabota*, Clapper *M. apiata*, Spike-heeled *Chersomanes albofasciata*, Pinkbilled *Spizocorys conirostris*, Stark's *Eremalauda starki*, Black-eared Finchlark *Eremopterix australis* and Grey-backed Finchlark *E. verticalis*) fall naturally into two groups: "red" and "grey". The "red" larks inhabit mainly the red sand dunes, while the "grey" larks inhabit the grey calcrete which borders the eastern banks of the two rivers, Nossob and Auob. Because of the very restricted area of the calcrete, it is not surprising that only three species are "grey" (Sabota Lark, Grey-backed Finchlark and Stark's Lark), the rest "red". Only when the populations of Grey-backed Finchlarks were so big as to exceed the carrying area of the calcrete did the birds begin to occupy the dunes; the same held true for the Lark-like Bunting *Emberiza impetuani* which is primarily a bird of stony habitats (Maclean 1970a). It is no coincidence that the seed-eating passerines mentioned here are also nomadic and gregarious at all times: Pink-billed Lark, Stark's Lark, the two finchlarks and the Lark-like Bunting. I therefore agree with Niethammer's conclusion, and with Willoughby (1969a) that the physiological benefits of pale coloration in a desert habitat are negligibly small.

The term "desert coloration" (Willoughby 1969a) is used for birds whose coloration matches that of the soil substrate on which the birds live, and is not confined to pale coloration. Birds in the Namib, which have "desert coloration" tend to be sedentary and to occur as endemic species or subspecies: Rüppell's Korhaan *Eupodotis rueppellii*, Double-banded Courser *Smutsornis africanus* (Fig. 2.3), Karoo Lark *Mirafra albescens*, Spike-heeled Lark *Chersomanes albofasciata*, Gray's Lark *Ammomanes grayi*, Tractrac Chat *Cercomela tractrac* and Karoo Chat *C. schlegelii*.

Fig. 2.3. Double-banded Courser *Smutsornis africanus* incubating its single egg on a cold morning (cf. Figs. 8.5 and 8.6)

Other ground-dwelling birds have "generalized cryptic coloration", are widespread beyond the borders of the Namib, have no forms endemic to the Namib and have few subspecies overall. The conclusion reached by Willoughby (1969a) from observations of Namib birds is again that desert coloration is protective against predation, in confirmation of Niethammer's (1959) earlier deductions, and that thermal advantages are probably insignificant.

Cloudsley-Thompson (1979) made the interesting observation, however, that pale colours may be cryptic on pale backgrounds, aposematic on dark backgrounds. Conversely, dark colours (especially black) are usually aposematic, but may be cryptic on dark backgrounds (black sands and lava flows). He also concluded that the thermal significance of colour is usually slight, if it exists at all. "The well-known rule that ground-nesting birds tend to have plumage colours and markings that conceal them from view when seen against the ground on which they live, is thus especially marked in deserts, where cover is sparse, and is well developed among larks in particular" (Cloudsley-Thompson 1979). It is therefore generally accepted that "desert coloration" is for crypsis.

Further support for these ideas comes from Thomas and Robin (1983) who described the neonatal plumage of Lichtenstein's Sandgrouse *Pterocles lichtensteini* as plain unpatterned brown above, slightly paler below. Only three other sandgrouse species (Spotted *P. senegallus*, Crowned *P. coronatus* and Indian *P. indicus*) have plain or nearly plain neonatal plumage. Since

Fig. 2.4. Dorsal view of chick of Namaqua Sandgrouse *Pterocles namaqua*, Namib Desert, Namibia

they are not closely related, with the exception of *P. lichtensteini* and *P. indicus* which may in fact be conspecific, and since all except *P. indicus* are adapted to extreme desert conditions, the loss of typical patterned plumage (Fig. 2.4) in these sandgrouse chicks has evolved at least twice in response to extreme aridity. Such a response has not occurred in the southern African deserts where selection for plain plumage appears not to have been so strong. It may also be that the southern African species are relatively recent "invaders" (Maclean 1984b) from species with patterned neonatal plumage in the African savannas.

3 Food and Energy

Most studies on the food and energy budgets of arid-zone birds have been done on seedeaters, probably because they are easier to study both in the field and in captivity. One of the few field studies on a non-granivorous species is that on the Inland (or Australian) Dotterel *Peltohyas australis*, though it was not quantified (Maclean 1976b). By day the Inland Dotterel (Fig. 3.1) is largely herbivorous, taking leaves, flowers and seeds of a variety of plants. The water content of the leaves of some of the plants is enough to satisfy the birds' water requirements, since this species was never seen to drink. The problem of osmoregulation when eating the saline leaves of plants like *Atriplex halocarpa* is solved by the activity of the salt glands which have secretory tubules of about 0.41 mm in length (Fig. 3.2), comparable to that of

Fig. 3.1. Inland Dotterel *Peltohyas australis* adult in breeding plumage, central Australian Desert, New South Wales

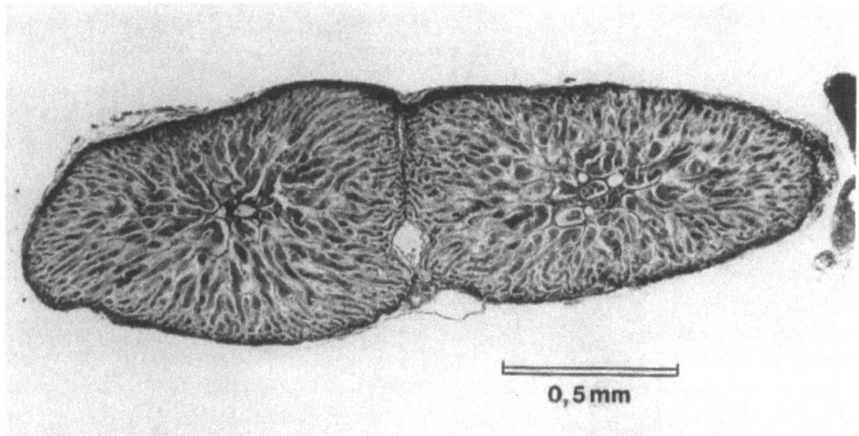

Fig. 3.2. Transverse section of the nasal (salt) gland of the Inland Dotterel, Australia

the tubules in marine charadriiforms (Staaland 1967). At night the Inland Dotterel takes mainly insects (beetles, ants and grasshoppers); this is its normal active period in the 24-h cycle. Insects may also be taken in daytime on mild mornings. The diet of this plover obviates any need to drink; this is an important aspect of the diets of many desert birds, even those which may be predominantly granivorous.

3.1 Seeds as a Food Resource

Relative to ants and mammals, experimental tests by Mares and Rosenzweig (1978) showed that birds are unimportant consumers of seeds in the Sonoran Desert of North America, but are more important than mammals in the Monte of northwestern Argentina. Birds have a high metabolic rate and, until the experimental source of seeds could be shown to be "reliable" (which was not the case because of the short duration of the experiments), birds would concentrate their activities elsewhere. These results do not necessarily show the true situation therefore. The need for a seed resource to be reliable is highlighted by the statements of Wiens and Johnston (1977): "Most seeds are small, and protected by a hard seed coat and, while frequently rich in calories and easily metabolized carbohydrates, are deficient in other essential nutrients, especially proteins. As a consequence, rather large quantities of seeds are required to satisfy the energetic and nutritional needs of individual birds." They add that the evolution of large-scale granivory in birds is especially prevalent in arid or semi-arid savannas. Plants in such regions produce large amounts of seed, but patchily in time and space. This in turn favours nomadic granivory because of the large variation in seed supplies.

Fig. 3.3. Karoo habitat near Nieu Bethesda, central Cape Province, South Africa

Seed densities in the Sonoran Desert of Arizona varied 78-fold in spatial distribution and 28-fold in temporal pattern (Reichman 1984). Densities were highest in depressions, whether natural or artificial, and lowest in dry washes and in open areas between shrubs. Differences in seed densities between microhabitats were greatest at times of high seed production and lowest in dry years. Small round seeds tended to form greater concentrations than large or long seeds.

Variations in seed density are correlated with (1) the location and productivity of the parent plants; (2) dispersal factors such as wind and water; (3) microtopographical features; and (4) foraging by granivorous animals which also act as seed dispersers.

Ants (mainly *Messor* spp.) are the main removers of seed in the Karoo (Fig. 3.3; annual rainfall at the two sites 180-200 mm/year), followed by nocturnal vertebrates, then diurnal vertebrates (Kerley 1991). This last category, which includes birds, removed insignificant amounts of seed from the experimental trays. Ants are the most important seed removers in summer, but nocturnal vertebrates become more important in winter. However, in this particular set of field experiments using field dishes, it was revealed that birds did not find or use the dishes, even though doves, canaries and sandgrouse were seen nearby. The suggestion is that desert birds rather rely on predictable, high-quality food patches and spend little time searching for new patches. The experimental dishes need to be left out for a much longer time than was the case (i.e. less than 4 days) in order to be discovered and used by birds. The results are therefore inconclusive with regard to seed removal by birds.

3.2 Small Passerines

Willoughby (1971) studied six species of larks in the Namib Desert. Their food preferences were shown by examination of stomach contents. On the diets shown in Table 3.1, these larks were independent of drinking water. The Red Lark *Mirafra burra* is confined to red dunes in the northwestern Cape. Its main food is large smooth seeds; these are augmented by succulent vegetation and some insects (Dean et al. 1991), which presumably allow this lark also to be independent of drinking water. The Dune Lark *Mirafra erythrochlamys* of the Namib Desert spends about two-thirds of its foraging time excavating in the sand for seeds, removing the seed-deficient surface layers to expose the seed-rich layers beneath (Cox 1983). The remaining third of the time is spent foraging for insects around the bases of the sparsely distributed grass clumps. It seems as if the intake of insects is necessary to supply the birds with water, since the water content of the seeds is insufficient for the birds' daily needs, especially when T_A is relatively high.

In a more detailed comparative study in the Negev Desert of Israel (Fig. 3.4), Shkedy and Safriel (1991) demonstrated that two lark species, the Crested *Galerida cristata* and the Desert *Ammomanes deserti* show different methods of exploitation of environmental and physiological resources. Crested Larks with a body mass of 39 g ($n = 118$) on average had a mean fat ratio of 12.9%, while the Desert Lark with a mean body mass of 22 g (n = 103) carried only 8.9% fat.

The Crested Lark is an opportunist, occurring in the desert only under favourable conditions. It feeds largely on seeds of 0.3-53.5 g mass and has a dielly variable fat ratio, but little seasonal variation in the fat ratio. The Desert Lark feeds on smaller seeds (mostly less than 1.0 mg in mass) and has similarly little seasonal fat variation, but has insignificant diel variation in fat, unlike the Crested Lark. The Desert Lark may store glycogen in the liver for overnight use. Fat storage in both species may be used to buffer variable foraging success from day to day; the greater storage capacity of the Crested Lark, in spite of its greater size and potentially better buffering against environmental vicissitudes, seems to indicate that it is less well able to "interpret" the desert environment than the more desert-adapted Desert Lark. The Crested Lark is a generalist which is more dependent on fat reserves than the specialized Desert Lark in an environment in which food reserves may fluctuate. The specialist is probably more efficient at obtaining its maintenance requirements in times of food scarcity, though in the case of the Desert Lark, its food supply of very small seeds is more abundant spatially and temporally by mass than are the larger seeds, so it may not face such an acute problem of food scarcity.

In their study of another small passerine, the Sociable Weaver *Philetairus socius* (Fig. 3.5), Williams and du Plessis (1995) found that its field metabolic rate (FMR), measured by using tritiated water, translates into about 48.7 kJ/day. The sample of birds weighed an average of 25.5 g. By comparison with the FMR of the House Sparrow *Passer domesticus*, that of the Sociable

Table 3.1. The diets of six species of larks (Alaudidae) in the Namib Desert, based on examination of stomach contents. (Willoughby 1971)

Species of lark	Percentage of food item in stomach, by volume		
	Seeds	Insects	Green vegetation
Sabota *Mirafra sabota*	60	40	Trace
Dune *Mirafra erythrochlamys*	32	68	Trace
Spike-heeled *Chersomanes albofasciata*	16	84	Trace
Gray's *Ammomanes grayi*	56	43	1
Stark's *Eremalauda starki*	77	19	4
Grey-backed Finch- *Eremopterix verticalis*	91	8	1

Fig. 3.4. Negev Desert, Israel (Photo by M.J. Lawes)

Weaver is about 60% of what might be expected, indicating considerable saving of energy. Mechanisms that may help to lower the energy output of these birds include roosting in the nest for about 2 h at midday, a reduced BMR, and huddling together with conspecifics. These adaptations, especially the lowered BMR, are directed towards desert conditions in which primary production is low. This will in turn lead to a reduced flux of water (4.6 ml/day), which is important since the birds do not normally drink, obtaining all their water from their food, either as preformed water or as metabolic water. The low water intake of the Sociable Weaver is only about 23% of the predicted value of 19.9 ml/day for a typical passerine weighing 25.5 g. Using data from other desert birds, a value of 8.92 ml/day was obtained, which is nearly twice that of the Sociable Weaver.

Fig. 3.5. Sociable Weaver *Philetairus socius*, Kalahari Desert, South Africa

The water-economy index (WEI) is the ratio of water intake to FMR (measured in ml/kJ), and is an indication of the amount of water used by an animal per unit of its metabolism, independent of the rate of energy intake. The average WEI for five species of desert birds (including the Phainopepla *Phainopepla nitens*, Gambel's Quail *Callipepla gambelii* and the Chukar Partridge *Alectoris chukar*) yielded a mean of 0.16 ml/kJ, whereas the Sociable Weaver's WEI averaged 0.09 ml/kJ. This last value is markedly lower than previously reported values. Clearly it is adaptive in a species with a low water intake. Because seeds contain little water, it is disadvantageous for a Sociable Weaver to eat more than about 50% of its food intake as seeds each day, and this is borne out by actual observations of stomach contents (Maclean 1973f).

3.3 The Ostrich *Struthio camelus* (Fig. 3.6)

Sauer and Sauer (1967) and Sauer (1971) claimed that the food of the Ostrich consists opportunistically of grass and other vegetation as well as small vertebrate and invertebrate animals. Water requirements are met partly by succulent food and, when necessary, also by drinking. However, Milton et al. (1993) observed that adult Ostriches eat only plant material, mostly soft annual herbs such as *Gazania*, *Indigofera*, *Lepidium* and grasses; the preferred food contains about 24% protein and 70% water, so that the birds in this study did not need to drink. Only Ostrich chicks up to the age of two months eat insects.

Fig. 3.6. Ostrich *Struthio camelus* male, South Africa

The Ostrich weighs between 70 and 100 kg as a rule; it is therefore nearly two orders of magnitude larger than other species of terrestrial bird for which data are available (Williams et al. 1993). The home ranges of wild Ostriches in the Namib averaged 84.3 km². Because of its size, the Ostrich has fewer evasive tactics open to it, such as seeking favourable microhabitats in the heat of the day. This study focused on three aims of the research into the biology of desert-adapted Ostriches: (1) to measure the field metabolic rate (FMR) and the water-influx rate (WIR); (2) to ascertain the diet of the birds and their sources of water; and (3) to quantify their daily activity patterns, from which an energy budget could be constructed. Its average field metabolic rate (FMR) is 18 040 kJ/day for adults (mean weight 88.25 kg), which is about 74% of the predicted value, and 15 428 kJ/day for subadults (mean weight 50.75 kg). The energy expenditure of adults was 26% lower than predicted, while subadults had an FMR close to that predicted. Adults may be better at foraging economically.

The Namib is hyperarid with scanty and localized rainfall (Fig. 3.7). Existence is challenged by a scarcity of water, relatively poor quality of food, intense solar radiation and frequent desiccating winds. The habitat in the Namib Desert consists of 84% open gravel plains and about 1% dry washes, yet the Ostriches spend about 65% of their day on the plains and 25.5% of the day in the washes. In this study their diet consisted of a narrow range of green plant material, the main species of which were *Monechma arenicola*, *Schmidtia kalahariensis*, *Blepharis* spp., *Trianthema triquetra* and *Dicoma capensis*. The mean water content of 12 species of plants was 57%, ranging

Fig. 3.7. Gravel plains in the Namib Desert, Namibia

from 36.4 to 69.7%. Although the abundance of the first two species listed was actually low, they made up about 40% of the Ostriches' diet. Many of the plants eaten were pulled up by the roots and swallowed whole; this strategy may not be open to herbivorous mammals because of the wear and tear by silica on their teeth.

Ostriches spent 7.5 h walking, accounting for 62.2% of their FMR; they roosted at night for 11.5 h, accounting for 19% of their FMR. Other activities accounted for less than 1% of the Ostriches' energy budget.

Compared with other non-passerine birds, Ostriches were economical of water, and had lower water-influx rates (WIRs) than predicted. Calculations of the water-economy index (WEI), which is the ratio of the WIR to FMR (in ml of water/kJ) and is independent of the rate of energy metabolism, showed that Ostriches conserve water (average of 0.17 ml/kJ) in the same way as do smaller birds of the desert, especially passerines (average for four species 0.16 ml/kJ). Adults did not drink during the study period, but subadults drank an average of 729 ml of water/day (about 22% of their daily water requirements), indicating that the water economy of adults is better than that of subadults.

3.4 Sandgrouse and Doves

The Black-bellied Sandgrouse *Pterocles orientalis* is a winter visitor to the Thar Desert of northwestern India. From the crops of 13 birds the minute seeds of *Indigofera linifolia* and *I. cordifolia* occurred in 92% and formed 21% by volume; one bird had an estimated 8700 *Indigofera* seeds in its crop (Christensen 1963). The larger seeds of the cultivated pulses *Phaseolus radiatus* and *P. aconitifolius* together formed 39% by volume, but occurred in only 30 and 23% of the crops respectively. The seeds of *Panicum* sp. and *Tephrosia purpurea*, both native plants, contributed jointly 28.6% of the crop contents, but occurred in only 30 and 15% of the crops respectively. No insect material was found in any of the crops. It appears that the winter distribution and numbers of Black-bellied Sandgrouse in the Thar Desert are influenced by prevailing food and water conditions.

The crop of one Chestnut-bellied Sandgrouse *Pterocles exustus* collected near Karachi, Pakistan, contained about 5600 seeds of *Indigofera cordifolia*, 51 seeds of *Tephrosia tenuis*, 89 seeds of *Indigofera uniflora* and 9 of *I. anabaptista*. That of a bird collected in Rajasthan, India, contained about 10 000 seeds of *Indigofera linifolia*, 350 of *Crotalaria* sp., 1 of *Tephrosia purpurea*, 5 of *Phaseolus radiatus* and about 100 unidentified seeds (Christensen et al. 1964).

Namaqua Sandgrouse *Pterocles namaqua* (Fig. 3.8) and Burchell's Sandgrouse *P. burchelli* (Fig. 3.9) in the Kalahari sandveld appear to feed exclusively on small dry seeds throughout their lives (Maclean 1968). Crop contents of adults and chicks contained nothing else; a chick less than a week old had 1400 tiny hard seeds in its crop. Among the most abundant seeds in the dunes, if only seasonally, are those of *Lophiocarpus burchelli* (Chenopodiaceae), an annual which grows commonly after rain. Other seeds in the crops of sandgrouse were small (< 2.5 mm diameter) and hard. However, Dixon (1978) showed that, although sandgrouse are mostly seedeaters, 49 crops of the Namaqua, Burchell's and Double-banded Sandgrouse *P. bicinctus* in Etosha National Park, Namibia, contained some insect remains (4% of crops) and some shells or fragments of terrestrial gastropods (10%) of crops. Insect remains included beetles and flies, while gastropods belonged to the family Achatinidae. It was concluded that these animal remains were ingested incidentally during feeding on seeds, except that the large number of shell fragments eaten by female Double-banded Sandgrouse during egglaying indicates that these birds had a specific calcium requirement for the formation of eggshells and that the shell fragments were taken deliberately. Dixon and Louw (1978) also showed that, in the Namib Desert Park, the Namaqua Sandgrouse feeds on the very small seeds of only seven plant species (*Tephrosia dregeana*, *Cleome diandra*, *C. luederitziana* and four unidentified) and very occasionally some green plant material (*Zygophyllum cylindrifolium* and *Sesuvium sesuvioides*). The measured values of energy and crude protein for the three identified seed species are very high for plant material, and they may also have a high oil content (Table 3.2).

Fig. 3.8. Namaqua Sandgrouse *Pterocles namaqua* male, Kalahari Desert, South Africa

Fig.3.9. Burchell's Sandgrouse *Pterocles burchelli* pair at waterhole, Kalahari Desert, South Africa

Table 3.2. Gross energy and protein content of seeds eaten by Namaqua Sandgrouse in the Namib Desert Park, with energy values for soya and starch for comparison. (Dixon and Louw 1978)

Species	Gross energy (kJ/g)	Crude protein (%)
Tephrosia dregeana	210.42	32
Cleome diandra	211.97	16
Cleome luederitziana	214.91	16
Soya beans	231.29	Not measured
Starch	177.24	Not measured

The food of the Namaqua Sandgrouse in the Namib varies very little in composition with the age of the bird and time of year, even when the adults are initiating breeding activity. The seeds on which the birds feed lie on the desert floor and are available all year. The standard (resting) metabolic rate (RMR) of the Namaqua Sandgrouse is calculated to be about 972 kJ/day.

Given a total flying time per day of about 4 h (involving a 12-fold increase in metabolic rate) and foraging time of 8 h (with a 3-fold increase in metabolic rate), the daily energy requirement of a sandgrouse weighing 182 g would be about 3380 kJ/day; this could be met by an intake of only 15.7 g of seed of *Cleome luederitziana* per day. Allowing for a digestion coefficient of 80%, the intake would need to be increased to 19.6 g seed/day. The relative abundance of the seeds of the food plants can be witnessed by even the most casual observer, so it is not surprising that the Namib can support huge numbers of Namaqua Sandgrouse, at least at certain times of the year.

Field studies by Hinsley and Ferns (1994) of breeding Black-bellied Sandgrouse *Pterocles orientalis* in Israel and Pin-tailed Sandgrouse *P. alchata* in Spain, augmented by observations on three other species (Spotted Sandgrouse *P. senegallus*, Crowned Sandgrouse *P. coronatus* and Double-banded Sandgrouse *P. bicinctus*) produced time and energy budgets for breeding adults at different stages, and for juveniles. Time budgets were made from direct observation. Each activity was assigned a cost and a daily energy expenditure (DEE) calculated as the sum of the costs of all activities (time in hours spent/activity per day × cost of activity in kJ/h) plus the cost of thermoregulation.

The time budgets of breeding sandgrouse of the two main species are shown in Tables 3.3, 3.4 and 3.5. Although males of the Black-bellied Sandgrouse spend about 15 h incubating, only 6 h are in daylight, while females spend all of their 9 h of incubation time during daylight. Males therefore have more time for foraging in daylight than do females, but they have to do so during the hottest time of the day. In order to balance the energy expenditure, females have to take in about 97 kJ/h, while males need to take in only about 79 kJ/h.

Male Pin-tailed Sandgrouse with young spend less time foraging (21%) than do females (32%), but more time standing alert (vigilance), since Montagu's Harriers *Circus pygargus* take young sandgrouse in Spain. When attacked, young flee into the nearest dense vegetation and the adults mob the predator. Male Pin-tailed Sandgrouse probably have to spend more time flying than do females, since they have to bring water to their young, but even with equal flight times, males need to take in more energy (62 kJ/h) than do females (44 kJ/h) because of their reduced foraging time. However, flight times for males with young are significantly greater than for females, since males may fly to water three times in the morning and again three times in the afternoon when their chicks are large (even though the water may be close by). When the young are large, males spend more energy in parental care than do females.

Table 3.3. Time budget of incubating Black-bellied Sandgrouse males and females, expressed as the percentage of daylight hours spent in each category of activity. (Hinsley and Ferns 1994)

Activity	Males	Females
Inactive	11	4
Foraging	32	24
Preening	9	5
Standing alert	1	1
Walking	4	4
Flying	4	4
Drinking	<1	<1
Incubating	39	58

Table 3.4. Time budget of Pin-tailed Sandgrouse males and females with young, expressed as the percentage of daylight hours in each category of activity. (Hinsley and Ferns 1994).

Activity	Males	Females
Inactive	30	31
Foraging	21	32
Preening	13	9
Standing alert	19	12
Walking	6	7
Flying	9	9
Drinking	<1	<1
Watering young	3	-

Table 3.5. Energy budgets of incubating Black-bellied Sandgrouse and of Pin-tailed Sandgrouse with young, expressed as the percentage of daily energy expenditure (DEE)for each activity. (Hinsley and Ferns 1994)

Activity	Black-bellied		Pin-tailed[a]		
	Males	Females	Males	Females	Juv.[b]
Night inactivity	-	20	16	15	18
Day inactivity	4	1	8	9	16
Foraging	21	17	12	18	24
Preening	5	3	6	4	2
Standing alert	<1	<1	8	5	<1
Walking	2	2	3	3	3
Flying	17	18	32	32	17
Day incubation	18	27	-	-	-
Night incubation	23	-	-	-	-
Thermoregulation	10	11	14	14	18
Total DEE (kJ)	380	348	221	225	176

[a] The value for "watering young" in Pin-tailed Sandgrouse was not included separately and is probably included in that for "standing alert".
[b] Juv., juveniles.

Contrary to published information, the mean masses of eggs of Black-bellied Sandgrouse (27.9 g; $n = 2$) and Pin-tailed Sandgrouse (19.6 g; $n = 12$) are close to predicted allometric values (28.8 g and 19.0 g respectively), so that the size of the eggs is not a means of reducing energy expenditure, as suggested by Thomas (1984b). The small clutch is probably related to the male's need to carry water to the young, rather than energetic parsimony by females.

The allocation of incubating duties between the sexes may be a compromise between camouflage by the female by day (they are harder to detect than males) and allowing the female enough time to feed in daylight. If males return late, the females do not wait for them, but leave the nest at the usual changeover time, at which time T_A is moderate. The option of departure in the event of a late-arriving female is not open to the male, since T_A is already high and the eggs may overheat. The cost of incubation is much greater for the male (156 kJ/day) than for the female (94 kJ/day), since the female is largely concerned with cooling the eggs by means of metabolically inexpensive methods (convective cooling when $T_A < T_B$, gular flutter and cutaneous evaporation). However, the cost of thermoregulating at night is relatively high for both sexes, so that their combined DEE is similar (179 kJ/day for males and 184 kJ/day for females). In addition to this there is the cost for females of flying to their nighttime roosts.

The Plumed (Spinifex) Pigeon *Geophaps plumifera* of arid Australia feeds on a wide variety of seeds of grasses, trees, shrubs and herbs. Insignificant amounts of green plant food are eaten, but insects (termites) may exceptionally account for up to 4% of food taken (Frith 1982). The most important seeds in one sample were those of *Triodia pungens* (25% by volume) and *Cleome viscosa* (23% by volume). Seeds of *Eucalyptus* gleaned from the ground may provide as much as 8% of the food. In other regions the seeds of *Acacia* replace those of *Eucalyptus* where the latter genus is scarce. Springs are effective drought refuges for a nucleus of Plumed Pigeons even in areas where rain has not fallen for 10 years, since the seeds and bulbs of plants that grow on the moist edges of springs assume importance in the birds' diet.

3.5 The Budgerigar *Melopsittacus undulatus*

The Budgerigar has a body-lipid content of about 4% in the wild, much the same as that of most small sedentary birds (this can build up to 25% in captivity; Wyndham 1980b, c, d). Its food is small seeds taken from the ground; they measure 0.5-2.5 mm long and have an energy content of about 18.9 kJ/g. Adults and young take the same food. It fills the crop only partially in the morning, but more completely in the afternoon (to about 8% of live mass), so that it can use the food in its crop as an energy source during the night when it does not use body fat; indeed it may even lay down some fat in the night. Under mild T_A Budgerigars do not need to drink and, unlike the Red-billed Quelea *Quelea quelea*, do not need to leave room in the crop for an evening drink. The function of body lipids in the Budgerigar may be to provide energy for flight, though the distances the bird covers are usually short. However, it performs aerial displays before roosting, when body lipids may be used as an energy source. Lipids conserved or laid down in the night are probably used as an energy source for the morning foraging flight.

3.6 Torpidity in Insectivorous Birds

In the Colorado Desert of California, at least some individuals of the Poorwill *Phalaenoptilus nuttallii*, a species of nightjar (Caprimulgidae), are known to enter a state of torpidity during the winter, instead of migrating to warmer latitudes. Jaeger (1949) found that the T_B of a bird in the field was a more or less steady 18-20 °C (close to T_A) and the bird became completely dormant, without perceptible heartbeat or breathing. Under experimental conditions, the T_B dropped to 5 °C at T_A 2-6 °C (Howell and Bartholomew 1959). An increase in the T_A was followed by a passive increase in T_B up to 15-19 °C, after which oxygen consumption increased rapidly and a T_B of 34 °C was quickly attained. Torpor might be maintained for days at a time and the Poorwill may become torpid every night when temperatures are low. It is not a constant feature of this species, however, and is unknown in other nightjars, so its exact cause in nature is unexplained. It must, however, constitute an energy-saving measure of significant importance.

4 Water (1): Drinking

Several studies have been done on drinking patterns and behaviour of birds in the field and in the laboratory. The main species involved are doves, sandgrouse and other obligate seedeaters, as well as some frugivorous/nectarivorous species such as mousebirds and honeyeaters. Although doves and sandgrouse have similar drinking patterns, their systematic relationship with one another is somewhat distant; they are products of convergent evolution (Maclean 1967a) and make for an interesting comparison with each other and with the equally convergent seedsnipes (Thinocoridae) of South America (Maclean 1969a), which have been shown to be charadriiforms (Sibley et al. 1968) most closely related to the Scolopacidae (Sibley and Monroe 1990).

4.1 General Field Observations

Studies of birds drinking at natural and man made waterholes in the Namib Desert (Fig. 4.1) allowed a division of the species into (1) regular; (2) occasional; and (3) "seldom" drinkers (Willoughby and Cade 1967). Regular drinkers drink daily and are water-dependent (although, surprisingly, some members of this category, like the Red-headed Finch *Amadina erythrocephala* (Fig. 4.2), the Lark-like Bunting *Emberiza impetuani* and the Black-throated Canary *Serinus atrogularis*, can survive long periods of water deprivation in the laboratory; the canary, however, lost mass steadily under deprived conditions). Occasional drinkers may drink when water is available, but appear not to be water-dependent. "Seldom" drinkers drink rarely or never, even when water is available to them.

Regular drinkers included the Namaqua Sandgrouse *Pterocles namaqua*, all pigeons and doves (Fig. 4.3), the Rosy-faced Lovebird *Agapornis roseicollis*, Red-eyed Bulbul *Pycnonotus nigricans*, Glossy Starling *Lamprotornis nitens*, Pale-winged Starling *Onychognathus nabouroup*, Cape White-eye *Zosterops pallidus*, Cape Sparrow *Passer melanurus*, Masked Weaver *Ploceus velatus*, Red-headed Finch, Common Waxbill *Estrilda astrild*, and all canaries and buntings. The Lark-like Bunting became independent of water after summer rains had resulted in a growth of green vegetation, in the sprouting of seeds and in the availability of small insects.

Fig. 4.1. Namaqua Sandgrouse *Pterocles namaqua* drinking at an artificial waterhole in the Kalahari Desert, South Africa

Fig. 4.2. Red-headed Finch *Amadina erythrocephala* drinking, Kalahari Desert, South Africa

Fig. 4.3. Cape Turtle Doves *Streptopelia capicola* drinking, Kalahari Desert, South Africa

Occasional drinkers included the Ostrich *Struthio camelus*, vultures, Pale Chanting Goshawk *Melierax canorus*, Red-faced Mousebird *Urocolius indicus*, Stark's Lark *Eremalauda starki*, Grey-backed Finchlark *Eremopterix verticalis*, Rock Martin *Hirundo fuligula*, Pied Crow *Corvus albus*, Sociable Weaver *Philetairus socius* and Scaly-feathered Finch *Sporopipes squamifrons*. Only about 10% of the population of the two lark species within an 8-km radius of the waterhole drank per day, and both species could survive without water on a diet of dry seeds. The Scaly-feathered Finch also showed an amazing ability to survive without drinking water and could be found foraging and nesting many kilometres from water in suitable bushy and grassy habitats.

Most bird species in the Namib remain relatively inactive during the hot hours of the day, usually seeking shade (Willoughby and Cade 1967). The Dune Lark *Mirafra erythrochlamys*, for example, forages until the sand reaches about 50 °C, then shelters in the tufts of *Stipagrostis sabulicola* until the sand temperature again drops below 50 °C when it emerges to forage in the late afternoon.

Columbids and sandgrouse are wary when coming to drink at exposed waterholes, and they gather in large numbers (Cade 1965). The behaviour of sandgrouse joining flocks at the waterhole is in many respects similar to that of other kinds of birds joining flocks to go to roost at night: (1) synchronized mass flights at set times of the day; (2) gathering at specific waterholes, even though there may be other suitable ones along the way; (3) gathering of groups of birds en route to the waterhole; (4) advertisement of the flight path

by loud calling; and (5) the joining of drinking flocks by breeding birds (Ward 1972). At Gobabeb in the Namib Desert one species of sandgrouse and four species of columbids regularly came to drink at a waterhole in the bed of the Kuiseb River (Cade 1965). The sandgrouse came from feeding plains about 20-30 km away, landed on a prominent rocky promontory about 500 m upriver until there were hundreds together, then flew down in small groups to the waterhole, landing in the riverbed about 3-25 m away, then walking cautiously to the water's edge. Drinking (and belly-soaking in the case of breeding males) took no more than 10-15 s, after which each bird flew directly away from the water with clattering wings, gaining height and disappearing towards the feeding grounds, usually without calling.

A small group of 7-9 Rock Pigeons *Columba guinea* came fairly regularly to drink, though not every day. They landed in the uppermost branches of a tall *Acacia* near the waterhole, sometimes staying there for half an hour before flying down to the perimeter of the pit in which the waterhole was located. There they might wait warily for several minutes before walking to the water's edge; other small birds arriving to drink might stimulate the pigeons to do likewise. After drinking the pigeons usually walked partway up the sloping edge of the pit before taking wing.

Laughing Doves *Streptopelia senegalensis* and Cape Turtle Doves *S. capicola* also perched in a tall tree on first arrival, usually before 10:00 and after 16:00 h. A vanguard might land near the water, but might take off before going to drink. After several such false starts, the doves flew down to the water in a steady stream until all had watered. Each bird drank in just a few seconds, then flew quickly away, sometimes colliding with one of the incoming birds as it did so. Namaqua Doves *Oena capensis* (Fig. 4.4) usually drank in the midday hours, landing on the ground in the shade of trees or bushes at the edge of the riverbed. There they sat motionless for a few minutes, then flew low over the ground to the water, drank quickly and flew away again.

Avian predators which took an interest in the drinking birds included Lanner Falcons *Falco biarmicus*, Gabar Goshawks *Micronisus gabar* and Pale Chanting Goshawks *Melierax canorus*. The Lanner Falcon stooped from a high perch in a tree or from circling flight to attack doves flying down to drink, but not always successfully. Although not many hunting attempts were observed, the conclusion may safely be drawn that the wary behaviour of birds coming to drink at exposed desert waterholes is an adaptive response to the presence of predators which gather at such places to prey on the drinkers. Gathering in large numbers also improves the combined ability of the drinking birds to detect predators.

"Birds are very wary at waterholes, and they seem very reluctant to lower their heads to drink" (Smyth and Coulombe 1971). Doves are probably especially vulnerable because, when they lower the head to drink, it has to be kept lowered for the full drinking time (at least 20 s in the case of the American Mourning Dove *Zenaidura macroura*; Fig. 4.5), and this may be why they are particularly wary at the drinking place. Gambel's Quail *Callipepla gambelii*, which drinks by tipping up every 3-5 s, takes much longer to drink and may spend up to 3 min drinking if uninterrupted. Most birds in this North American study drank around mid-morning (09:00-11:00 h), usually after feeding. Granivorous birds were all water-dependent. Birds on other diets were mostly independent of water, though some might drink at T_A above about 40 °C.

In 20 dawn-to-dusk watches at waterholes (both natural and artificial) in different parts of Australia it was revealed that 118 species of about 35 families drank at least sometimes (Fisher et al. 1972). All observed members of the families Dromaiidae, Glareolidae, Columbidae, Psittacidae (includes Cacatuidae), Hirundinidae, Meliphagidae, Estrildidae, Grallinidae, Ptilonorhynchidae and Corvidae (a total of 47 species) were dependent on water for drinking. About 60% of the observed species were independent of water, compared with about 70% of Namib species (Willoughby and Cade 1967), but this difference may not be significant, taking into account local differences in the physical characteristics of the two study areas (the Namib is cooler on average).

Numbers of certain species drinking in one day could be very great: 47 000 Budgerigars *Melopsittacus undulatus*, 17 750 Zebra Finches *Poephila guttata*, 4500 Cockatiels *Nymphicus hollandicus*, 1750 Spiny-cheeked Honeyeaters *Acanthagenys rufogularis*, 1558 White-plumed Honeyeaters *Meliphaga penicillata* and over 1000 each of Little Corellas *Cacatua sanguinea* (Fig. 4.6) and Galahs *C. roseicapilla*.

Fig. 4.5. American Mourning Dove *Zenaidura macroura*, Sonoran Desert, Arizona, USA

Fig. 4.6. Little Corella *Cacatua sanguinea*, New South Wales, Australia

Three general patterns of drinking times show up in the water-dependent species: (1) drinking confined to two short periods in the early morning and late afternoon; (2) similar to (1), but the two periods are longer and less well separated by a complete absence of midday drinking; and (3) drinking throughout the day with no peaks, or else a small increase in drinking at midday. The only diurnal species drinking before and after sunrise was Bourke's Parrot *Neophema bourkii*: the birds began to arrive an hour before sunrise, most drank at dawn and a few after sunset. Wyndham (1980a) showed that, although the Budgerigar feeds mostly in the early morning and late afternoon, it drinks mainly between mid-morning and mid-afternoon, but he suggested that this species drinks only infrequently, unless T_A is very high: his experimental Budgerigars survived without water at T_A below 20 °C, but not at higher T_A.

Species with two peaks of drinking activity per day may consist of some individuals which drink twice a day (the evening peak is usually lower than the morning peak, except in the Common Bronzewing *Phaps chalcoptera*), but most individuals drink only once a day, or maybe even once every 2-3 days. The frequency of drinking increases with rising T_A, and some species may not drink at all at moderate to low T_A.

Why the Meliphagidae (honeyeaters) should be water-dependent is not known. It may be related to their high level of activity, since they tend to be pugnacious, chasing each other continually during feeding. Nectar-feeders in Africa (sunbirds) and North America (hummingbirds) drink infrequently if at all. Of the 52 species of insectivorous birds in this study, only 9 were water-dependent (Australian Pratincole *Stiltia isabella* (Fig. 4.7), Welcome Swallow *Hirundo neoxena*, Tree Martin *Hetrochelidon nigricans*, Fairy Martin *P. ariel*, Crimson Chat *Ephthianura tricolor*, Orange Chat *E. aurifrons*, Magpielark *Grallina cyanoleuca*, Apostlebird *Struthidea cinerea* and White-browed Woodswallow *Artamus superciliosus*). In the case of the aerial-feeding hirundinids, high activity may again be the reason for the need to drink. The pratincoles may have been breeding at the time of the study and incurred extra water demands. Explanations for drinking in the other species are not offered.

Although only 40% of observed species are water-dependent, 58% of all individuals belong to this group, most of which are granivorous and fairly large (e.g. cockatoos and pigeons): "the total biomass of seed-eating birds must be very much greater than that of insect-eaters, the majority of which are very small" (Fisher et al. 1972). Seed-eaters are also most abundant in the North American deserts.

Only columbids, the Australian Bustard *Ardeotis australis*, the Budgerigar and some estrildids drink by immersing the bill and sucking without raising the head to swallow. All species approach the waterhole warily, even the Emu *Dromaius novaehollandiae* (Fig. 4.8) and birds of prey, which might seem not to have any predators other than humans. Budgerigars often settle in the middle of a pool, float and drink for only about 2-4 s before taking off again. Avian predators (accipiters and falcons) are most common at waterholes where birds drink in greatest numbers (Fisher et al. 1972).

Fig.4.7. Australian Pratincole *Stiltia isabella* pair at nest, New South Wales, Australia

Fig. 4.8. Emu *Dromaius novaehollandiae* male and young drinking, New South Wales, Australia

4.2 Doves

The American Mourning Dove (body weight about 150 g) is common throughout the deserts of the American southwest. Although most abundant near water, its excellent powers of flight allow it to "exploit areas that are many miles from available water" (Bartholomew and Dawson 1954). It does not seem to need to drink more than once a day even in very hot weather, visiting the waterholes morning and evening only. Mourning Doves can drink in 10 min an amount of water equivalent to 17% of their own body weight. At moderate T_As they can go for 4-5 days without drinking water or feeding on succulent plant food. As long as they have water to drink, they can maintain their body weight at high T_As, unlike humans and most domestic mammals. The availability of water is particularly important in summer when the birds are cooling by evaporation, laying eggs and feeding their young on crop milk.

Under experimental conditions, Mourning Doves deprived of water at T_A 39 °C lost about 15% of their body weight, but this was regained within a few minutes of being provided with drinking water. At such a high T_A the doves drank four times as much water as at 23 °C (probably indicating a high EWL for cooling purposes).

Adaptations of the Mourning Dove to an arid environment include: (1) a capacity to endure prolonged high T_B and dehydration without damage to tissues; (2) a capacity to make up water deficits rapidly when water becomes available; and (3) good powers of flight which enable it to exploit food sources far from water.

The Ground Dove *Columbina passerina* inhabits a hot, dry climate in California. Unlike some other species of doves (e.g. the American Mourning Dove) it does not gather at regular times of the day to drink, but seems to do so at intervals throughout the day when T_A is high (Willoughby 1966). It was not observed to drink in winter under natural conditions, even though water was available all the time in irrigation canals and reservoirs.

Diamond Doves *Geopelia cuneata* in northwestern Australia began to forage at dawn and went to drink between 10:00 and 14:00 h, in company with other species of birds, such as Zebra Finches and Budgerigars (Schleucher 1993). They never drank more than once a day, yet Crested Pigeons *Ocyphaps lophotes* regularly drank twice a day. Because as many as five raptors hunted at the waterhole during these hot midday hours, the Diamond Doves, Zebra Finches and Budgerigars often stayed in the trees for 3-4 h and drank only briefly, but the raptors were inactive between about 12:00 and 14:00 h, during which time the other birds could drink in safety.

Immediately after drinking the Diamond Doves returned to the bush along the dry riverbed to begin feeding again. Feeding continued even when the T_A was maximal between 14:00 and 15:00, but intensified towards evening. Other ground-foraging bird species did not feed at this time of day. Diamond Doves and Plumed Pigeons *Geophaps plumifera* were often exposed to direct sunlight in the heat of the day and experienced T_A above

40 °C for as much as 10 h of the day. They might rest on the ground for an hour at a time, where the temperature was as high as 62-65 °C, without showing any signs of distress. Diamond Doves were never entirely inactive during the daylight hours (Schleucher 1993).

It is not always clear why doves of different species drink at different times of the day, including throughout the day in some cases, or why some species gather to drink in very large numbers, while others arrive singly or in small groups. These different drinking patterns may be related to predation or some other environmental factor, but they seem not to be necessarily dependent on temperature regimes.

4.3 Sandgrouse

The 16 species of sandgrouse (family Pteroclidae) constitute a suborder of the order Charadriiformes (Sibley and Monroe 1990). They are one of two charadriiform families to have become exclusively phytophagous, the other being the seedsnipes (family Thinocoridae) of South America (Maclean 1969). The seedsnipes (Fig. 4.9) are somewhat unusual in that they do not drink under normal natural conditions, having access to succulent plant material from which they obtain all their water requirements (Maclean 1969). Twelve of the 16 species of sandgrouse drink in the morning 2-4 h after sunrise; some individuals drink again in the afternoon (Ward 1972). The remaining four species (*P. bicinctus*, *P. quadricinctus*, *P. indicus* and *P. lichtensteini*) drink mainly after sundown and some may drink again just before dawn. In spite of the large size of most drinking flocks, sandgrouse forage, roost and nest on the ground in relatively small, scattered groups or solitarily, probably for increased crypsis. Because of this, it is possible that the drinking places have become the main information centres. If, for example, a bird is finding insufficient food in its foraging area, it can join a different group of birds at a waterhole and fly with them to their (presumably better) feeding grounds.

The Black-bellied Sandgrouse *Pterocles orientalis* of the Thar Desert of northwestern India drinks from shortly after sunrise, assembling in thousands on dry barren flats near the waterhole ("tank") (Christensen 1963). Once flocks have gathered, they rise and circle the waterhole before landing, drinking and flying to their feeding grounds. Up to 8000 birds may water at one time, but it is said that the flocks used to be bigger than this.

In India and Pakistan the Chestnut-bellied Sandgrouse *Pterocles exustus* may fly up to 16 km/day to water, the first birds usually gathering from about an hour after sunrise. About 1 or 2 h later, the flocks rise, circle over the waterhole and, if undisturbed, land to drink, sometimes right in the water. Very few individuals water again at sunset (Christensen et al. 1964).

Fig. 4.9. Least Seedsnipe *Thinocorus rumicivorus* male at nest, Atacama Desert, Chile

Circumstantial evidence shows that Namaqua Sandgrouse *Pterocles namaqua* (Fig. 3.8) and Burchell's Sandgrouse *P. burchelli* (Fig. 3.9) in the Kalahari sandveld fly up to 80 km one-way to drink at waterholes in the morning (Maclean 1968). When travelling such long distances, they sometimes fly at heights of 300 m above the ground, at a cruising speed of about 70 km/h. Drinking times are fairly specific, from 1-2 h after sunrise in summer and about 2-2.5 h after sunrise in winter. Burchell's Sandgrouse usually drinks on average later than the Namaqua Sandgrouse. A few Namaqua Sandgrouse drink again in the late afternoon about 1.5 h before sunset. Main drinking times of the four species of southern African sandgrouse in the Makgadikgadi Botswana, differ as shown in Table 4.1 region, (Ginn 1977). As in the case of doves, mentioned earlier in this chapter, it is not always clear why different species have different drinking programmes, in particular why some drink by day and others by night.

Flocks of Namaqua Sandgrouse usually gather on high ground near the waterhole until hundreds or even thousands of birds are present; then they fly up and land a few metres from the water, run to the water, drink and take off from the water's edge (Maclean 1968). Burchell's Sandgrouse land right at the water, or even on the surface where they float like ducks, drink quickly and fly straight away. They can take off from a floating position on the water without difficulty.

Sandgrouse do not drink water with a NaCl content more than 40-50% of that of seawater (Maclean 1968; Table 4.2).

Table 4.1. Main drinking times of four species of sandgrouse in the Makgadikgadi region of Botswana. (Ginn 1977)

Species	Drinking times
Double-banded Sandgrouse *Pterocles bicinctus*	Evening after sunset
Burchell's Sandgrouse *P. burchelli*	Morning (from about 09:30)
Yellow-throated Sandgrouse *P. gutturalis*	Afternoon (about 15:30)
Namaqua Sandgrouse *P. namaqua*	Morning (from about 09:30)

Table 4.2. Salinities of some waterholes in the Kalahari Gemsbok National Park, compared with tap water and seawater, showing sandgrouse drinking preferences. (Maclean 1968)

Source of water	Molarity		Drunk by sandgrouse
	Na^+	Cl^-	
Tap water	0.50	0.17	
Seawater	470.00	448.00	
St. John's Dam	0.40	0.13	++++
Houmoed	30.00	7.10	+++
Rooiputs	42.00	15.00	+++
St. John's Well	58.00	34.00	++
Monro	64.00	38.00	++
Kasper se Draai	168.00	108.00	+
Kameelsleep	250.00	174.00	+
Kij Kij	272.00	197.00	-
Kransbrak	824.00	665.00	-

+, Water drunk; -, water not drunk; the most favoured waters have the largest numbers of (+) signs.

Both Namaqua Sandgrouse and Burchell's Sandgrouse drink by immersing the bill briefly, sucking, then raising the head to swallow (Cade et al. 1966). Namaqua Sandgrouse take an average of 9.6 such draughts (males) or 9.4 draughts (females) per drinking attempt, while Burchell's Sandgrouse take an average of 7.4 and 7.0 draughts for males and females respectively. Each drinking attempt takes no more than 5-10 s. Individuals drink from 5-15 ml of water, depending on the degree of thirst. Water-deprived sandgrouse drink more than others and take 1.1-2.0 ml/draught. The amount of water which Namaqua Sandgrouse can drink and store in their crops varies between 22.5 and 29.5 ml, which represents 13.7-16.4% of the birds' body mass respectively (Willoughby and Cade 1967).

4.4 Other Species

Mousebirds (family Coliidae; Fig. 4.10) in the Kuiseb Valley at Gobabeb in the Namib Desert feed on the leaves, flowers and fruits (green or ripe) of *Salvadora persica*, preferring ripe fruit when it is available (Cade and Greenwald 1966a). The water content of this fruit is about 66% of the fruit's

Fig. 4.10. White-backed Mousebird *Colius colius*, South Africa. (Photo by G.K. Lestrange)

wet mass, which allows the birds to be independent of drinking water. Mousebirds drink except when *Salvadora* is in full fruit and/or when T_A is moderate to low. The two species observed (Red-faced *Urocolius indicus* and White-backed *Colius colius*; Fig. 4.10) drink in the same way as doves do, immersing the bill and sucking the water up without needing to raise the head to swallow.

Larks of the Namib Desert, which feed partly or largely on insects do not drink. These larks are the Sabota Lark *Mirafra sabota*, Dune Lark *M. erythrochlamys*, Spike-heeled Lark *Chersomanes albofasciata* and Gray's Lark *Ammomanes grayi* (Willoughby 1971). The largely seed-eating Stark's Lark *Eremalauda starki* and Grey-backed Finchlark *Eremopterix verticalis* drink, but only about 10% of the population of Stark's Lark appear to drink each day, and it seems as if large flocks of Grey-backed Finchlarks live far from water and may not need to drink.

4.5 Conclusions

The drinking patterns of birds differ in several basic respects: (1) time of day; (2) frequency; (3) gregariously or not; and (4) whether by sucking or by tipping up after each mouthful. The time of day when birds visit water may be governed partly by the activity patterns of predators (especially raptorial birds) and partly by daily temperature regimes, in that the drinking birds will attempt to avoid those times of day when predators are most likely to be active at the waterholes, and when T_A is low, so as to avoid flying during the hotter hours. The social patterns of drinkers may also be dictated in part by the enhanced detection of predators when the drinkers are in large flocks together, but this is clearly not the whole story. The need to drink may be avoided altogether by the inclusion in the diet of some insect food or some succulent vegetation, whether leaves, fruit or flowers. In the absence of such foods, the species of birds concerned may need to drink. Drinking by sucking allows the birds to take in more water per unit time, but carries with it the disadvantage of having the head lowered throughout the bout of drinking, thereby reducing the ability of the bird to detect predators. It is therefore surely no coincidence that birds which drink by sucking (doves, sandgrouse, mousebirds and some estrildids) are gregarious, a condition which enhances the detection of predators by the presence of many pairs of eyes and/or the posting of sentries at the waterhole.

5 Water (2): Water Regulation

5.1 Introduction

Chapter 4 dealt with the acquisition of water by drinking and, to some extent, also by taking in preformed water with the food, which may obviate drinking partly or completely. Once the water has been gained by the bird's body, it has to be conserved as far as possible. However, some water has to be lost during processes of excretion and evaporative heat loss; such losses in a water-poor environment, such as the arid regions of the world, have to be kept to the barest minimum compatible with survival. The account which follows deals with the single most important function of adaptation to the arid environment, and that is the conservation of water.

The subject will be dealt with by examining the control of thirst, the sources and quality of available water, the structure and function of the avian kidney, nasal (salt) gland and hindgut, and evaporative water loss (EWL). Comparative studies of arid-adapted vs non-arid-adapted birds will be reviewed to determine to what degree they differ. Finally a selection of case histories of desert birds will be presented and a set of conclusions drawn.

"Terrestrial life is an aqueous phenomenon" (Bartholomew and Cade 1963). One could say that all life is an aqueous phenomenon, since life is inseparable from water (Bartholomew 1972). Water constitutes 55-75% of the adult avian body mass (Takei et al. 1989). Birds gain water through drinking ("free" water) and from food, either directly as "preformed" water or indirectly by catabolism (metabolic water). Water is lost via urine, faeces, respiratory and cutaneous evaporation, through the nasal glands, and (in breeding females) in the eggs. Birds have a high tolerance to dehydration; they may lose 30-50% of their body mass as water before they die. Rehydration is rapid: birds can consume 12-18% of their body mass in water when given access to it after dehydration (Dawson and Bartholomew 1968).

Birds are not as good as mammals at water conservation, so that arid-zone birds may be particularly hard pressed to manage physiologically in an environment that is poor in water by definition. Furthermore, a parsimonious water economy is denied to birds because, unlike most small rodents, they are diurnal (with few exceptions) and they do not live in burrows, so they lack the advantages of reduced temperatures and increased relative humidities provided by these structures (Dawson and Bartholomew 1968). They are also obliged to have a high water-turnover rate because of their high rate of pulmocutaneous water loss (a consequence of their high metabolic rate) and their limited capacity for concentrating electrolytes in the urine.

Water balance is maintained if the water gained through drinking (*A*), preformed water in the food (*B*) and metabolic water (*C*) equal that lost through evaporation (*D*), renal and cloacal excretion (*E*) and extrarenal excretion via the salt gland, if present (*F*), so that (Skadhauge 1981)

$$A + B + C = D + E + F.$$

Because granivorous birds obtain little water in their food, when evaporative cooling takes place, water intake needs to be augmented by drinking. Birds feeding on animal food seldom need to drink, even in deserts.

The amount of water drunk is proportional to the evaporative water loss, which in turn is inversely proportional to the body mass of the bird: the smaller the bird, the higher the evaporative water loss and the greater the proportional amount of water drunk (Fig. 5.1).

The ratio of water lost by evaporation to water produced is related to body mass, but mass-specific oxygen consumption is about one-third greater in passerines than in other orders of birds (Bartholomew 1972). It has been shown that the metabolic production of water can never equal the loss of

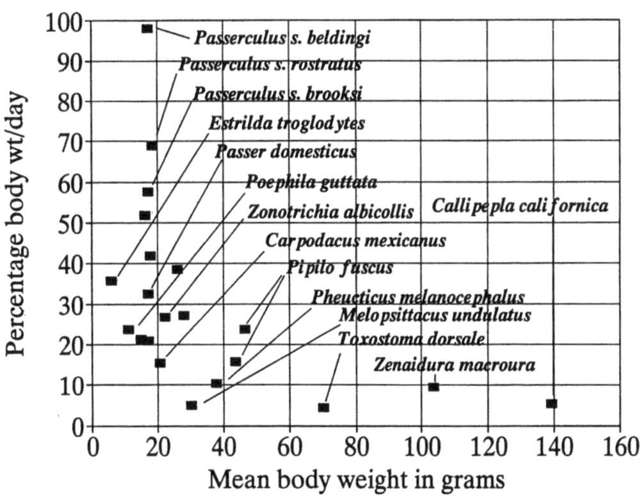

Fig. 5.1. Voluntary water intake (expressed as the percentage body weight/day) relative to body weight of some selected bird species; evaporative water loss is proportional to the percentage intake of water. (Redrawn from Dawson and Bartholomew 1968)

water by evaporation, even in the zone of thermoneutrality, so that metabolic water must always be augmented by the intake of preformed or free water. This would seem to be especially necessary in small birds. However, several species of small birds have been found to survive on a diet of air-dried grain in the absence of drinking water, although not all of these are birds of arid zones (Table 5.1). When dehydrated, some of the species of birds listed in Table 5.1 can reduce EWL, but how they do it is not known.

Table 5.1. Birds that can survive on a diet of air-dried grain without drinking water. (Bartholomew 1972)

*Budgerigar *Melopsittacus undulatus*
Horned Lark *Eremophila alpestris*
*Stark's Lark *Eremalauda starki*
*Grey-backed Finchlark *Eremopterix verticalis*
*Scaly-feathered Finch *Sporopipes squamifrons*
*Red-headed Finch *Amadina erythrocephala*
Cut-throat Finch *Amadina fasciata*
*White-throated Silverbill *Lonchura malabarica*
*Zebra Finch *Poephila guttata*
*Lark-like Bunting *Emberiza impetuani*
Savannah Sparrow *Passerculus sandwichensis*
*Black-throated Sparrow *Amphispiza bilineata*
Brewer's Sparrow *Spizella breweri*
Vesper Sparrow *Pooecetes gramineus*

*, Species of arid habitats.

One way in which water loss can be reduced is by reducing metabolic rate through decreased activity. Most birds also reduce their intake of food under dehydrated conditions; this may also have the effect of reducing urinary-faecal water loss. However, even at reduced levels of metabolism, EWL far exceeds metabolic water production.

As has been shown in Chapter 4, the ecological drinking patterns of desert birds fall into three categories (Bartholomew and Cade 1963; Cade 1964). Drinking in desert birds is largely confined to those that (1) live near surface water; or (2) have powers of flight that allow them to travel relatively long distances to water (Dawson and Bartholomew 1968). Birds at a watering place are vulnerable to predation and survey the area carefully before going to drink. The drinking time is reduced by sucking or by bill-tilting very quickly, e.g. in doves, sandgrouse, mousebirds and some estrildids.

Reliance on metabolic water as the main source of water would require one or more of the following water-conservation mechanisms (Calder 1964):

1. Reduced activity and/or metabolic rate to minimize heat production and water loss accompanying greater respiratory demands;
2. Tolerance of limited hyperthermia as an alternative to evaporative cooling;
3. Thin plumage to facilitate heat loss at T_A below T_B (this is no doubt a function of birds being diurnal, as opposed to small fossorial mammals, which are nocturnal and have a thick pelage);

4. Modification of pulmonary anatomy, movements or physiology to reduce exposure of moist surfaces or decrease water flux across these surfaces;
5. Increased renal resorption of water to produce more concentrated urine;
6. Cloacal resorption of water.

Of these mechanisms, reduced activity at high T_A (but not reduced metabolic rate), tolerance of hyperthermia (increase in T_B from 41 °C to 43.5 °C), thin plumage for greater conductance (the reciprocal of insulation), reduction of evaporative water loss if deprived of water, and increased renal and/or cloacal resorption have all been recorded, for example, in the Australian Zebra Finch *Poephila guttata*, and are all adaptations (or preadaptations) to the arid environment.

Some species of birds which can live indefinitely without drinking on a diet of air-dried seeds include the Budgerigar *Melopsittacus undulatus*, Zebra Finch, Scaly-feathered Finch *Sporopipes squamifrons*, Black-throated Sparrow *Amphispiza bilineata*, Savannah Sparrow *Passerculus sandwichensis rostratus*, Lark-like Bunting *Emberiza impetuani*, Red-headed Finch *Amadina erythrocephala* (Fig. 4.2) and Violet-eared Waxbill *Granatina granatina* (Willoughby 1968). These species are exceptional in their ability to eat dry food and to do without regular drinking.

5.2 Thirst

Thirst is regulated by three mechanisms: (1) cellular dehydration; (2) extracellular dehydration; and (3) the renin–angiotensin system. The dipsogenic effect of angiotensin II (ANGII) is well known, but work on cellular and extracellular water is relatively recent. "Intake of NaCl is as important as intake of water in avian osmoregulation..." (Takei et al. 1989). Drinking can be induced in the Feral Pigeon (or Rock Dove) *Columba livia* and Japanese Quail *Coturnix japonica* by injecting hypertonic NaCl (which causes cellular dehydration), even when the birds are replete with water. The same effect can be obtained by injecting hypertonic sucrose or mannitol, but not with glucose or urea. Osmosensitive receptors in the pigeon lie in the preoptic and lateral hypothalamus of the ventral part of the brain, much as in mammals.

Dipsogenic volume receptors which monitor extracellular fluid volume appear to be located in the extravascular interstitial section of the extracellular compartment, at least in the Feral Pigeon. (Since the tonicity of the cells and the pressure of the blood–vascular system have to be maintained in a relatively constant condition, it is the interstitial fluid that is subject to the greatest volume changes, so that the volume receptors are most logically located in this compartment.) Withdrawal of blood from the Feral Pigeon and the Japanese Quail induced drinking; the same effect may be obtained by injecting 50% polyethylene glycol. Thirst is induced in two ways: (1) cellular or osmotic thirst is caused by an increase in the osmotic pressure of the extracellular fluid; and (2) volume or extracellular thirst is caused by a decrease in the volume of the extracellular fluid (Takei and Kobayashi 1980).

As mentioned above, thirst can also be stimulated by ANGII acting as a chemical stimulus. ANGII is dipsogenic in most birds, but weakly so in birds that do not normally drink in the wild (for example, the Shikra *Accipiter badius* and the Spotted Owl *Athene brama*) and in species adapted to arid regions, such as the Budgerigar and the White-throated Silverbill (or Three-coloured Munia) *Lonchura malabarica*. Takei and Kobayashi (1980) suggest that response to ANGII may have become attenuated or lost in certain species of birds, but that an angiotensin thirst mechanism has evolved as an appropriate adaptive response in vertebrates generally.

5.3 Water Sources and Water Quality

5.3.1 Preformed Water

Since relatively few desert birds have access to free water for drinking, and since most of them must augment their water intake with preformed water, they must get it from succulent food, whether plant or animal. Insectivorous passerines seldom need to drink, but they take in a high concentration of amino acids (insect haemolymph has 50-100 times the concentration of amino acids as that of mammalian blood). Since birds excrete uric acid with relatively little loss of water, the excretion of nitrogenous wastes is not seriously expensive of water.

Similarly, raptors obtain most of their water from their food and seldom need to drink, even when using evaporative cooling for several hours. Several species of falconid occur in totally waterless regions, e.g. Peregrine Falcon *Falco peregrinus* (Middle East), Sooty Falcon *F. concolor* (Sahara), Aplomado Falcon *F. femoralis* (American southwest), Saker Falcon *F. cherrug*, some populations of Lanner Falcon *F. biarmicus* (northern and southern Africa), the American Kestrel *F. sparverius* and the Pygmy Falcon *Polihierax semitorquatus* (eastern and southern Africa).

5.3.2 Tolerance of Saline Water

Birds subjected experimentally to saline water (NaCl solutions) respond in one of three different ways:

1. Some increase their water intake with increasing salinity, until a critical concentration is reached, at which they cease drinking (American Mourning Dove *Zenaidura macroura* and House Finch *Carpodacus mexicanus* are examples).
2. Some drink an amount equivalent to the amount of fresh water that would be drunk, even if they lose body mass in the process (examples are the California Quail *Callipepla californica* and the Budgerigar).
3. The third pattern is found in birds inhabiting salt marshes (various North American sparrows; Dawson and Bartholomew 1968).

Fig. 5.2. White-winged Dove *Zenaida asiatica*, Sonoran Desert, Arizona, USA

The ability of birds to discriminate between distilled water and saline waters varies. American Mourning Doves, White-winged Doves *Zenaida asiatica* (Fig. 5.2) and Inca Doves *Columbina (Scardafella) inca* fail to distinguish between distilled water and 12.5% seawater, but do distinguish between salinities of higher concentrations than this, invariably preferring the lower concentration when a choice is available. The California Quail will drink distilled water, 12.5% seawater and 25% seawater as readily as each other, but at higher concentrations will drink the more dilute solution if available. Hydrated California Quails will not drink 37.5% seawater, but will drink 70% seawater if dehydrated.

Most land birds that have been tested cannot maintain body mass on solutions stronger than 0.3 M NaCl or 50% seawater (Dawson and Bartholomew 1968). These birds include such arid-adapted species as the Budgerigar and the House Finch. Rather better performers include the Blackthroated Sparrow and one subspecies of the Song Sparrow, namely *Melospiza melodia cooperi*, which penetrates to desert watercourses in California. Among desert birds, those showing the greatest ability to obtain physiologically useful water from salt solutions are the Rock Parrot *Neophema petrophila* and the Zebra Finch of Australia. The Rock Parrot appears to be able to gain weight while drinking 75-100% seawater.

Fig. 5.3. Stark's Lark *Eremalauda starki*, Kalahari Desert, South Africa

Some individuals of Stark's Lark *Eremalauda starki* (Fig. 5.3) and the Grey-backed Finchlark *Eremopterix verticalis* could maintain their body mass when drinking NaCl solutions of up to 0.4 M, but for most of the birds 0.3 M was near the upper limit of tolerance. This performance is not unusual, so these two lark species appear not to have any peculiar ability to withstand high salt concentrations in their drinking water (Willoughby 1968). Stark's Larks and Grey-backed Finchlarks lost mass rapidly for about four days when deprived of water; the loss of mass then slowed down suddenly and the birds gained mass to a new low level at which body mass remained more or less constant on a diet of dry millet seed and no water (Willoughby 1968). Even when provided with distilled water, the water consumption of both Grey-backed Finchlark and Stark's Lark was below the expected volume (8% and 13% of body mass per day, respectively). *Ad libitum* water consumption by most birds of comparable size (18 g) exceeds 20% of the body mass per day.

The ability of certain salt-marsh passerines to concentrate electrolytes in the urine is astonishing. Some performances of birds in this regard are shown in Table 5.2. The ability of birds to concentrate electrolytes renally appears to lie more in the larger number of nephrons with loops of Henle than in the length of the loops themselves.

Table 5.2. Mean maximum Cl^- (mEq/l) and mean maximum osmotic (mOsmol/l) concentrations of urine of some tested land birds. (Dawson and Bartholomew 1968)

Species	Mean maximum Cl^-(mEq/l)	Mean maximum osmotic concentration (mOsmol/l)
Savannah Sparrow *Passerculus sandwichensis beldingi*	960	2000
Savannah Sparrow *P. s. brooksi*	527	1000
Black-throated Sparrow *Amphispiza bilineata*	510	-
Rock Wren *Salpinctes obsoletus*	372	-
House Finch *Carpodacus mexicanus*	370	850
American Mourning Dove *Zenaidura macroura*	189	-
Rock Dove *Columba livia*	-	298

Passerines adapted to salt marshes in North America are theoretically preadapted to inhabiting deserts, so that it is of interest to note that some Savannah Sparrows *Passerculus sandwichensis*, a polytypic species of which some subspecies are salt-marsh-adapted, have successfully occupied waterless desert islands off northwestern Mexico (Dawson and Bartholomew 1968). Two subspecies of Savannah Sparrow, *P. s. beldingi* and *P. s. rostratus* can maintain body mass on undiluted seawater (equivalent to 0.5 M NaCl) and solutions of up to 0.55 M NaCl. *P. s. beldingi* can drink up to 0.7 M NaCl and can tolerate serum Cl^- levels of up to 215 mEq/l (normal serum Cl^- level of birds = about 135 mEq/l) and a serum osmolarity of 490-610 mOsmol/l (normal serum osmolarity of birds = about 390 mOsmol/l) (Poulson and Bartholomew 1962) and urine osmolalities of up to 577 mmol/kg (Goldstein et al. 1990). Other passerines and upland populations of the Savannah Sparrow cannot tolerate serum Cl^- levels or serum osmotic pressures much above normal. *P. s. beldingi* also has higher urine flow rates and larger kidneys than upland subspecies of the Savannah Sparrow, as well as "substantial volumes" of hyperosmotic fluids (up to 811 mmol/kg) in the lower intestine; it also has uric acid extending up into the ileum (Goldstein et al. 1990). The urine concentration of this salt-marsh inhabitant can be as much as 2.7 times greater than plasma osmolality. The high rate of urine flow (7-10 µl/min, as against 1.7 µl/min for upland populations) and refluxing of hyperosmotic urinary fluids into the intestine are said to reflect the unusual circumstance of a salt-marsh dweller without salt glands (Goldstein et al. 1990).

The species mentioned so far as having high tolerance for seawater do not have salt-secreting glands, so their ability to excrete electrolytes must depend on renal function, assisted no doubt by water resorption in the hindgut and cloaca.

5.4 Pathways of Water Loss

5.4.1 Excretion

5.4.1.1 The Avian Kidney

The kidneys of birds, located in depressions in the pelvis, are each made up of three lobes: cranial, medial and caudal (Sturkie 1976). Each lobe consists of a number of lobules drained by a common ureteral branch. The avian nephrons are of two main types: (1) the reptilian type of nephron (RT) lacking loops of Henle and situated in the cortex; and (2) the mammalian type (MT) with long or short loops of Henle located in the medulla. The avian kidney is relatively larger than that of mammals: the ratio of kidney mass to body mass ranges from 6-21 mg/g and is inversely proportional to the size of the bird. The relative size of the kidneys is also inversely proportional to increasing aridity of habitat: Thomas and Robin (1977) have shown that sandgrouse (*Pterocles*) kidneys are relatively small in mass and have long collecting ducts, which indicate good salt-conservation abilities; sandgrouse species in drier habitats have relatively lighter kidneys. Sandgrouse species in Morocco can be ranked in order of decreasing relative kidney mass and increasing arid-adaptability thus: *P. alchata* > *P. orientalis* > *P. coronatus* > *P. senegallus* > *P. lichtensteini* and the same ranking can be made with regard to increasing water-transport capacity of their belly plumage, a subject which will be dealt with more fully in Chapter 8.

Light kidneys appear therefore to be related to overall water conservation in birds. If so, the relatively small number of glomeruli per unit body mass can be regarded as equivalent to permanent and adaptive glomerular antidiuresis. At the same time, the long collecting ducts are functionally suited to tubular antidiuresis (as happens in mammals also). A substantial proportion of the water in ureteral urine may be removed by post-renal resorption in the colon, "especially in birds on a low-salt diet" (Thomas and Robin 1977).

Birds produce uric acid ($C_5H_4O_3N_4$) as their nitrogenous waste product (Schmidt-Nielsen 1964), i.e. they are uricotelic. It has low solubility, is crystalline in form and can be excreted as a semi-solid paste with a low water content. Even so, urine from the kidneys has to be quite fluid in order to pass through the thin ureters, so that resorption of water must occur in the cloaca in order to produce the semisolid end product found in birds (Schmidt-Nielsen et al. 1963). This probably occurs by the active resorption of ions from the fluid, followed by the passive withdrawal of water. Greater concentrating powers of the kidney are probably not feasible, since they would precipitate too much uric acid in the kidney tubules, which would lead to kidney malfunction. The resorption of cations from the cloaca in order to resorb water necessitates a mechanism whereby these cations can in turn be removed from the system; the nasal salt gland performs this function (see below).

If the ureters of chickens are connected to the exterior, bypassing the cloaca, they lose excessive amounts of Na^+, because cloacal resorption is bypassed. Extrarenal secretion of ions is therefore probably a prerequisite of avian evolution in highly terrestrial habitats, and a further prerequisite is the excretion of uric acid as the nitrogenous waste product. The production of uric acid as the main excretory end product in birds is an evolutionary consequence of oviparity: the ancestral reptile embryo in its cleidoic egg had to evolve a non-toxic nitrogenous waste product which could be stored in the allantoic cavity until hatching time. The capacity of uric acid to be precipitated out of a fluid solvent adds to its preadaptiveness as the avian excretory product in desert habitats.

Furthermore, uric acid has a lower hydrogen content (one H per N) than urea (CH_4ON_2 = two H's per N), so that the amount of water produced during oxidation of proteins in birds is higher than in mammals: 1 g of protein yields 0.396 g H_2O when urea is produced and 0.499 g H_2O when uric acid is produced, an increase of 26%. Most metabolic water in birds comes from carbohydrates and fats. However, the amounts of metabolic (oxidative) water produced from all sources are insignificant relative to the amounts of water needed for thermoregulation. Even under unstressed conditions, the evaporative water loss exceeds metabolic water production approximately twofold. This does not even include the loss of water in faeces and urine. This is why birds generally have to augment their water intake by drinking and/or by feeding on succulent foods.

Uricotely (the excretion of uric acid as the principal nitrogenous waste product) provides birds with a mechanism for conserving water: molecule for molecule, twice as much nitrogen is excreted by uric acid as by urea; while mammals lose about 3.8 ml of water per millimole of urea excreted, birds lose less than 0.3 ml (about 8% of 3.8) to get rid of a similar amount of nitrogen as uric acid (Dawson and Bartholomew 1968). Uric acid is also osmotically much less active than urea. It cannot, however, be regarded as an adaptation to life in the desert, but is rather a preadaptation – a legacy from the archosaurian reptile ancestor.

Birds fare less well in the excretion of electrolytes in the kidneys, since they have poor concentrating ability because of their structure: the loop of Henle is less well developed than that in the mammal (Dawson and Bartholomew 1968). Many of the nephrons in the avian kidney lack loops of Henle altogether, resembling more the nephrons of reptiles. Birds kept on salt water show an increase in urine osmolarity when the blood becomes hypertonic. The volume of urine does not change, but the Na^+ and Cl^- concentrations increase. The output of other substances remains unchanged. Although birds are largely uricotelic, about 30% of the nitrogenous waste is in the form of ammonium ions (NH+). Generally the kidneys of birds secrete far less Na^+ than those of mammals (Peaker and Linzell 1975). Furthermore, although medullary tissue in the kidneys of desert-dwelling Desert Sparrows *Passer simplex* and Saxaul Sparrows *P. ammodendri* is more elongated and more complex in structure than in the kidneys of the mesic-zone Eurasian Tree Sparrow *P. montanus* the concentrating ability of the more complex

Table 5.3. Rate of urine flow (kg/min) in hydrated and dehydrated Domestic Fowl *Gallus domesticus* and Budgerigar *Melopsittacus undulatus.* (Sturkie 1976)

Species	Condition	Mean urine flow (kg/min)
Domestic Fowl *Gallus gallus*	Hydrated	0.298
	Dehydrated	0.018
Budgerigar *Melopsittacus undulatus*	Hydrated	0.105
	Dehydrated	0.028

kidney appears to be no better: the main interspecific differences in osmoregulatory function between these species seems therefore to lie in the lower intestine, rather than in the kidneys (Amanova 1984).

Birds can regulate the rate of urine flow (secretion) by the kidney depending on the state of hydration. Examples are given in Table 5.3 for the Domestic Fowl *Gallus gallus* and the Budgerigar. Interestingly, although only the Budgerigar is arid-adapted, the performances of the two species are very similar when dehydrated, the Domestic Fowl being rather the better performer in reducing its urine flow; however, the hydrated fowl allows a far greater flow of urine through the system than does the hydrated Budgerigar.

Table 5.4. Renal concentrating ability of selected bird species. (Skadhauge 1981)

Species	Osmotic ratio, urine: plasma	mOsmol/l	Osmotic and ionic concentration(mEq/l) Na^+	Cl^-
Dehydrated				
Domestic Fowl *Gallus (gallus) domesticus*	1.6	538	134	517
Budgerigar *Melopsittacus undulatus*	2.3	848	-	-
Roadrunner *Geococcyx californianus*	-	593	-	-
Zebra Finch *Poephila guttata*	2.8	1005	-	-
Galah *Cacatua roseicapilla*	2.5	982	-	-
Laughing Dove *Streptopelia senegalensis*	1.7	661	-	-
Crested Pigeon *Ocyphaps lophotes*	1.8	655	-	-
Emu *Dromaius novaehollandiae*	1.4	459	-	-
Kookaburra *Dacelo gigas*	2.7	944	-	-
Ostrich *Struthio camelus*	2.7	800	-	-
Salt-loaded				
Savannah Sparrow *Passerculus sandwichensis beldingi*	5.8	2000	-	960
Savannah Sparrow *P. s. brooksi*	3.2	1000	-	527
House Finch *Carpodacus mexicanus*	2.4	850	-	370
Zebra Finch *Poephila guttata*	1.4-2.8	1027	-	224
Domestic Fowl *Gallus (gallus) domesticus*	1.1	362	164	-
Gambel's Quail *Callipepla gambelii*	-	669 - 962	470 - 493	363
Australian Pipit *Anthus novaeseelandiae*	3.0	-	50	-
Blackthroated Sparrow *Amphispiza bilineata*	-	-	-	703
Rock Wren *Salpinctes obsoletus*	-	-	-	403
Sage Sparrow *Amphispiza belli*	1.9	674	-	345

In birds with salt glands (see Sect. 5.4.1.2) 60-85% of the NaCl is eliminated by the glands, the rest by the kidneys (Sturkie 1976). Na^+ is the principal osmotically active electrolyte in both plasma and urine. It is actively absorbed in the intestine and transported to the kidney where it may be resorbed into the plasma or excreted by the kidney. The renal concentrating mechanism depends on the size, number and orientation of the loops of Henle, which form a countercurrent concentrating mechanism. The presence of salt glands is associated with increased renal mass in non-passerine birds.

Renal concentrating ability varies with the state of dehydration and/or salt-loading. A concentrating ability of greater than 1000 mOsmol/l (except in the salt-marsh populations of the Savannah Sparrow) has not been found in any bird; this is a urine:plasma ratio of 3 (Table 5.4; Skadhauge 1981).

5.4.1.2 The Nasal (Salt) Gland

The salt gland provides an extrarenal pathway for the maintenance of homeostasis since bird kidneys are generally incapable of excreting all the ingested electrolytes (Holmes and Phillips 1985). In terrestrial (and marine) environments where available drinking water is hyperosmotic to the body fluids, salt glands perform this function. The size and function of the gland depend on environmental conditions; secretion occurs only immediately after ingestion of hyperosmotic fluids and ceases once the body contains enough osmotically free water to maintain homeostasis. Osmoreceptors are probably located in the blood vessels of the heart. Secretion is controlled by both nervous and hormonal stimuli.

Most of the work done up to the early 1970s was on ducks and gulls; so little has been done on land birds, let alone desert birds (Peaker and Linzell 1975), that this is a wide open field for future investigation. The nasal glands are situated in or around the orbit (Fig. 5.4). Supraorbital glands lie in a depression in the skull above each orbit and may cover the surface of the skull to meet in the midline in certain marine birds (Common Sandpiper *Tringa hypoleucos*, Knot *Calidris canutus*, Little Auk (Dovekie) *Alle alle*). The glands are served by blood vessels passing through a supraorbital foramen. Each gland has two ducts passing into the nasal cavity, so that the secretion flows out of the nostrils; in birds without nostrils (gannets, cormorants) the secretion flows through the internal nares along the roof of the mouth to the tip of the bill.

Each gland may consist of 2-25 lobes (Fig. 3.2), depending on the species, each surrounded by connective tissue. Radially arranged secretory tubules open into the central canal (collecting duct) of the lobe. These canals join up to form secondary ducts which in turn enter primary ducts which run into the nasal cavity.

The gland secretes almost exclusively NaCl in solution, without any trace of mucus.

The blood supply to the nasal gland arises from the internal carotid artery and enters the orbit via a foramen situated somewhat dorsal to the optic

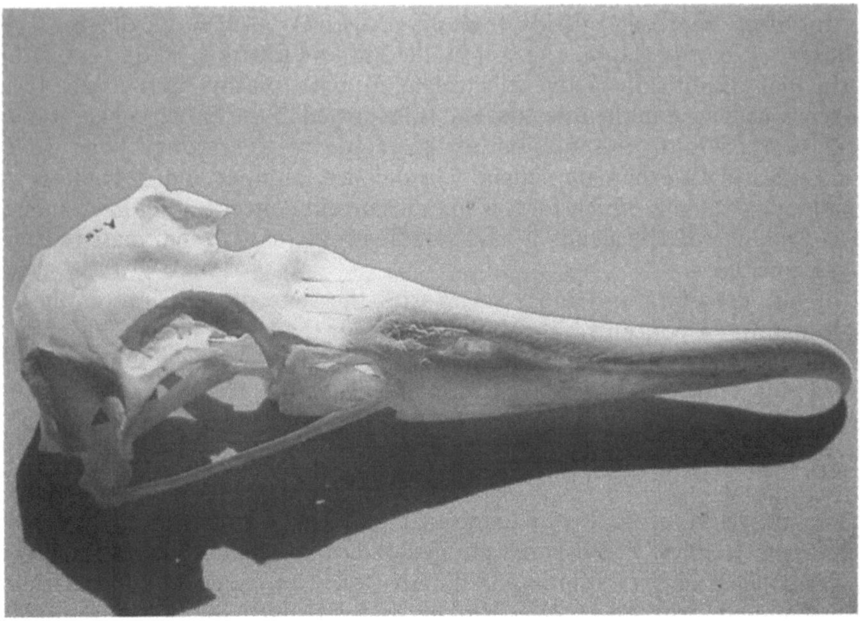

Fig. 5.4. Skull of Wandering Albatross *Diomedea exulans* showing positions of supraorbital nasal (salt) glands above each orbit

nerve (the internal ophthalmic artery). Branches arise from this artery to the gland, the artery then proceeding to the bill anteriorly. Nerve fibres from a ganglion in cranial nerve V pass through the orbit to innervate the nasal gland. This ganglion receives fibres also from nerves VII and IX.

Secretion by the nasal gland can be stimulated by introduction of seawater, other salt solutions or hypertonic sucrose by stomach tube or intravenously (Peaker and Linzell 1975). In the Herring Gull *Larus argentatus* secretion begins 1-5 min after intravenous salt loading, but it is inhibited by general anaesthesia. Stimulation of the secretory nerve in the orbit evokes secretion. Early workers concluded that secretion was initiated cholinergically, since acetylcholine and methacholine (= acetyl-β-methylcholine) stimulate secretion and atropine blocks it. There is still no evidence of a functional innervation of the gland by adrenergic fibres.

Initially, the view was that osmoreceptors initiated secretion by the nasal gland, but, since blood volume increases after administration of hypertonic NaCl solutions, it was suggested that volume or stretch receptors might be responsible instead. Birds deprived of water undergo a decrease in blood volume and an increase in blood tonicity (plasma osmolarity and NaCl concentration); this is followed by nasal-gland secretion, which seems to indicate that osmoreceptors and not volume receptors may be responsible for initiating secretion. However, long-term water deprivation inhibits secretion, possibly through mediation by the sympathetic nervous system acting to constrict the glandular blood vessels.

It is now clear that an increase in blood volume is not the primary stimulus for nasal-gland secretion. Hyperosmotic NaCl, sucrose and mannitol evoke secretion, but hyperosmotic urea, glucose and KCl_2 do not. Stimulation is not merely an increase in plasma osmolarity, but in tonicity which involves net water movement out of a cell in response to an osmotic gradient. Plasma tonicity is increased by the introduction to the blood of solutes which do not pass easily through the cell membrane, thereby resulting in cell shrinkage as water passes out of the cell. The conclusion is that the receptors are probably tonicity receptors (also known as osmoreceptors). These receptors appear to be located either in the heart or major arteries; they send stimuli to the brain via the vagus nerves.

The concentration of NaCl in the secretion of the nasal gland is a function of the salinity of the drinking water and the kind of food eaten. For example, birds that eat marine invertebrates take in more salt than those that eat vertebrates, since the invertebrates are often in osmotic equilibrium with seawater (Table 5.5).

Functional salt glands have been demonstrated in ten (and possibly 11) avian orders (Thomas and Phillips 1978). These include Struthioniformes, Pelecaniformes, Procellariiformes, Sphenisciformes, Gruiformes, Charadriiformes, Galliformes (only the Sand Partridge *Ammoperdix heyi*), Anseriformes, Phoenicopteriformes, Falconiformes and probably also Gaviiformes. Salt glands may be important in the excretion of NaCl resorbed along with water in the cloaca and colon. Because secretion by the salt glands is stimulated by an intake of NaCl and other electrolytes, as well as osmotically active solutes of other kinds (e.g. sucrose, mannitol), it is generally accepted that osmoreceptors play an important role in the initiation of salt-gland function. Dehydration can also induce secretion: the Sand Partridge and the Ostrich *Struthio camelus* will secrete also under conditions of thirst or high T_A. There is evidence that a degree of antagonism exists between the osmoreceptors (promoting secretion in response to increasing osmolarity of body fluids, i.e. the solute concentration determined by the depression of freezing point) and volume receptors (inhibiting secretion in response to falling volumes of body fluids).

The flow of blood to the salt gland is under nervous control, specifically that of the sphenopalatine nerve, stimulation of which causes vasodilatation. Hormonal control by adrenalin, noradrenalin or stimulation of the cervical sympathetic nervous system causes vasoconstriction and a reducti0on in secretion.

Table 5.5. Concentrations of nasal-gland secretions of some seabirds in relation to their usual food supply. (Peaker and Linzell 1975)

Species	Food	Concentration of nasal-gland secretion (mmol)
Cormorants	Mainly fish	500
Herring Gull *Larus argentatus*	Vertebrates and invertebrates	600-800
Leach's Petrel *Oceanodroma leucorrhoa*	Invertebrates	900-1100

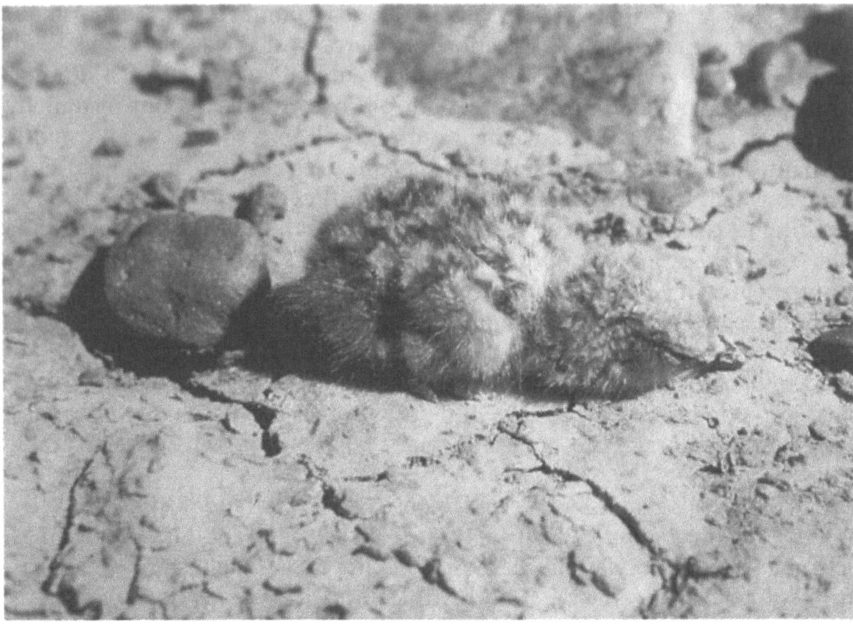

Fig. 5.5. Chick of Australian Pratincole *Stiltia isabella*, New South Wales, Australia; the white deposit around the nostrils is salt secreted by the supraorbital salt glands

A study by Staaland (1967) concerned only marine and aquatic Charadriiformes (Scolopacidae, Charadriidae, Laridae and Alcidae), but showed that the concentrating ability of the avian nasal (salt) gland is positively correlated with the length of the secretory tubules which, along with the size of the gland, was correlated with the birds' ecology. The lengths of the secretory tubules varied from about 0.2 mm in the freshwater species of wader to 0.4 mm in the marine waders and most gulls. The length of the secretory tubules of the nasal (salt) gland of the chicks of the Australian Pratincole *Stiltia isabella* (Fig. 5.5) averaged 0.147 mm (Jesson and Maclean 1976). In the adult Inland (Australian) Dotterel *Peltohyas australis* the tubules measured about 0.41 mm (Fig. 3.2; Maclean 1976b), comparable to the tubule length in marine charadriiforms (Staaland 1967). The pratincole and dotterel both breed in the arid interior of Australia, the former during the first half of summer as a rule (Maclean 1976a).

Stimulation of the nasal glands to secrete is not necessarily osmotic, but results at least partly from the process of eating the prey (Cade and Greenwald 1966b). These workers showed that, within 9 min after starting to eat a mouse, a Gabar Goshawk *Micronisus gabar* started to secrete fluid into its external nares (Cade and Greenwald 1966b). About half of the total of 10-15 μl of secretion accumulated during the meal and the rest within 5 min of finishing the mouse. The eating of raw chicken meat usually did not elicit any secretion, but an intact sparrow carcass stimulated much the same amount of secretion as a mouse. The predominant ions in the nasal secretion

Fig. 5.6. Roadrunner *Geococcyx californianus*, a large terrestrial cuckoo adapted to arid conditions in North America. (Photo by Olaf Wirminghaus)

were Na^+ and Cl^- and were well above the concentration of the plasma even at their lowest levels. Concentrations in the nasal secretion of 450-500 mmol/l for both Na^+ and Cl^- were considered usual, despite some experimental inaccuracy as a result of salt crystals in the nares from previous secretions. Concentrations of these two ions in the blood were 160 mmol/l for Na^+ and 127 mmol/l for Cl^-, more or less as to be expected.

Sixteen accipitrids and five falconids were found to secrete fluid from a nasal gland. In those studied, the glands lie partly within the orbit and partly within the orbital sinus. The cloacal fluid of falconiforms is usually hypotonic to the plasma, yet the birds are mostly independent of water. Their prey has a water and electrolyte composition similar to their own, but EWL from the birds' bodies means that they incur a salt load from their food. It is probable that the nasal glands play a role in eliminating excess electrolytes (Cade and Greenwald 1966b).

The Ostrich has large nasal glands which can secrete high concentrations of K^+, Na^+, Ca^{2+} and Cl^-. Nasal salt secretion was found in the Desert Partridge and the Ostrich when exposed to high T_A in the absence of osmotic stimulation, and the salt gland of the Desert Partridge also responded to salt loading (the Ostrich was not tested) (Schmidt-Nielsen et al. 1963; see also Pinshow et al. 1983; Thomas et al. 1984). When exposed to high T_A as well as being deprived of water, the Ostrich secreted fluid with high concentrations of Na^+, K^+, Ca^{2+} and Cl^-, but the proportions of these ions varied. The

Fig. 5.7. Australian Pipit *Anthus novaeseelandiae* thermoregulating: standing in shade, increasing surface area for heat loss by spreading wrists away from body, and reducing insulation by sleeking plumage

Roadrunner *Geococcyx californianus* (Fig. 5.6) also has a nasal gland, but it is much larger in the young than in the adult.

The domestic duck can secrete a fluid with a NaCl concentration about three times that of blood plasma. Control of fluid volume and of fluid concentration appear to be independent or at least separate functions (Inoue 1963). Flamingos can secrete NaCl at high concentrations via their salt glands when given a salt load (Peaker and Linzell 1975).

Although not a desert bird, the Snowy Plover *Charadrius alexandrinus*, a small wader (30-40 g), breeds on the Great Salt Plains of North America where thermal and osmotic stresses are high (Purdue and Haines 1977). Birds in captivity could maintain body mass at salinities of drinking water of 0.20-0.25 M NaCl, but not at 0.3 M (seawater is 0.6 M NaCl). The birds maintained their mass on a diet of mealworms and no water. The conclusion is that Snowy Plovers do not have an exceptional tolerance to salinity or water deprivation when compared with other bird species. The salt gland is moderately well developed and has a salt-secreting potential intermediate between that of a marine bird and that of a terrestrial plover like the Killdeer *C. vociferus* of North America. The Snowy Plover's insect diet, coupled with its water-conservation behaviour (e.g. standing in pools of water in hot weather) emancipate it from the need to drink the saline waters of its salt-plain habitat. "To survive in these stressful habitats, the birds eat succulent foods and depend on appropriate maintenance behavior" (Purdue and Haines 1977).

However, "few birds that live in deserts have salt glands and it should not be imagined that their presence and ability to secrete hypertonic salt solutions is necessary for survival under such conditions" (Peaker and Linzell 1975). Furthermore, some birds lacking salt glands can drink salt water and even seawater without ill effect. For example salt-loading shows that the Australian Pipit *Anthus novaeseelandiae* (Fig. 5.7) can excrete urine with NaCl ions up to a concentration of 570 mEq/l. The nasal glands are unimportant in the excretion of electrolytes (Rounsevell 1970), so that cloacal resorption of water is likely to be operative in the water economy of this species.

5.4.1.3 Cloaca and Rectum

Sodium and water are withdrawn from the urine in the lower end of the alimentary canal (cloaca, rectum and possibly also the caeca, when present), precipitating the relatively insoluble uric acid and leaving more water to be resorbed. The relationship between the functions of the cloaca, rectum, kidneys and salt glands has not been fully worked out (Dawson and Bartholomew 1968). Hydrated Budgerigars, weighing about 30 g, lose about 1 g (about 3% of body mass) of water per day through urine and faeces which contain about 75-80% water; in the dehydrated condition they lose only about 0.65 g of water per day. Faecal water in the Zebra Finch, weighing 10-13 g, also contains about 80% water when drinking water is freely available; cloacal water loss is 1.28 g/day (about 10-13% of body mass), but when dehydrated the faecal water content decreases to 65% and water loss via urine and faeces decreases to 0.54 g (4-5% of body mass), so the Zebra Finch is almost as good as Budgerigars at water conservation, despite its smaller body size (for further details, see Table 5.6).

Table 5.6. Water content of faeces and urine, and daily water loss by some small desert birds. (Dawson and Bartholomew 1968)

Species	Mean body mass (g)	Faecal water (%)		Daily water loss (% body mass)	
		Hydrated	Dehydrated	Hydrated	Dehydrated
Budgerigar	30	75-80	60	1 g (3.3%)	0.65 g (2.2%)
Galah[a]		83	66	-	-
Zebra Finch	10-13	80	65	1.28 g (10-13%)	0.54 g (4-5%)
Black-throated Sparrow	13.5	81	57	-	-
Sage Sparrow[a]		76	62	-	-
Brewer's Sparrow[a]		80	59	-	-
Stark's Lark	18.6[b]	74	52	-	-
Grey-backed Finchlark	17	75	51	-	-

[a] Skadhauge (1981)
[b] Maclean (1993)

Fig. 5.8. Flock of Galahs *Cacatua roseicapilla*, New South Wales, Australia

Absorption of Na^+, K^+ and Cl^- follows the same pattern in the Galah *Cacatua roseicapilla* (Fig. 5.8) as in the Domestic Fowl: the higher the intraluminal concentrations, the higher the absorption rates (Skadhauge 1974). Water flow is large and solute-linked, which implies that the cloacal transport parameters are such that the dehydrated Galah can let urine into the cloaca at the rate of about 100 µl/kg/h with a NaCl concentration of 50-100 mM and a K^+ concentration of 100-400 mEq/l without further loss of water. Wild-caught Galahs had coprodeal contents with an osmolarity of around 700 mOsmol/l and ionic concentrations of around 50 mEq/l (Skadhauge 1974). This seems to allow a resorption of about 70% of the NaCl in the ureteral excretion, and possibly also some slight resorption of water.

Cloacal water loss is a linear function of food intake in the Barbary Dove *Streptopelia risoria* and can therefore be reduced by reducing food intake. This cannot be a permanent solution to the problem of cloacal water loss, however, and will be complicated by the water content of the food and the degree of water resorption in the gut (Bartholomew 1972). The dietary water index (I_{dw}) is a function of the amount of food taken in, the amount of excreta emitted and their respective water contents:

$$I_{dw} = \frac{\text{mass of food} \times \text{\% water in food}}{\text{mass of excreta} \times \text{water in excreta}}$$

Table 5.7. I_{dw} for two species of larks weighing about 17 g on a diet of air-dried seeds and no water. (Willoughby 1968)

Measurements	Stark's Lark *Eremalauda starki*	Greybacked Finchlark *Eremopterix verticalis*
Food eaten/day (g)	1.90	2.20
Excreta emitted/day (g)	0.34	0.38
Water in food (%)	9.00	9.00
Mean water in excreta (%)	52.00	51.00
Dietary water index (I_{dw})	0.97	1.02

When $I_{dw} = 1$, water is neither gained nor lost. When $I_{dw} < 1$, eating increases the rate of water loss and when $I_{dw} > 1$, eating results in a net gain of water. Thus the index can be controlled by varying the quality of the food, the degree of water resorption and the concentration of the urine. Dehydration generally results in a diminution of the percentage of water in the urine. With *ad libitum* water most birds have a faecal/urinary water content of about 80%, which can be reduced to about 50% when dehydrated. Nevertheless, some small birds can maintain an I_{dw} of close to 1 on a diet of air-dried seeds (Table 5.7).

The problem of separating kidney function from cloacal function is difficult to solve, but a pattern emerges with respect to these functions (Bartholomew 1972):

1. Bird species which can survive without drinking water can obtain water from more concentrated solutions of NaCl than can obligate drinkers.
2. The capacity to survive without drinking is positively correlated with the number and length of loops of Henle in the kidneys, as in mammals.
3. The structure of the kidney and the ability to survive without drinking are correlated with the capacity to produce urine with a high concentration of electrolytes.

However, to date, the precise site of water resorption in the gut of birds has not been demonstrated; though urine moves freely from the cloaca to the rectum (and caeca) in the Domestic Fowl and the Roadrunner (Fig. 5.6), it has not been shown that water is resorbed there. The Roadrunner has ineffective kidneys, but possesses a salt gland. Those passerines and the Budgerigar which produce dry faeces and can maintain mass without drinking must resorb water somewhere in the lower intestine (hindgut), since they do not have extrarenal excretory mechanisms.

5.4.2 Evaporative Water Loss (EWL)

Evaporative water loss increases far more markedly at T_A above thermoneutrality than does oxygen consumption (and the consequent production of metabolic water), thereby exacerbating the problem of maintaining water balance (Bartholomew 1972). Below the lower critical temperature (T_{LC}) oxygen consumption increases, but EWL remains almost unchanged; the smaller the bird the higher its T_{LC}, and the greater its mass-specific conductance, thus favouring its adaptation to high T_A. Indeed, birds of tropical to subtropical deserts generally have high T_{LC} (Bartholomew 1972). For a diurnal bird of low-latitude arid zones a small body size with high conductance and a high T_{LC} is therefore likely to be selected for as an adaptation to high daytime T_A (Bartholomew 1972). This may account for the large numbers of small birds in arid zones generally.

In three species of North American quails (California Quail, Gambel's Quail *Callipepla gambelii* and Northern Bobwhite *Colinus virginianus*), weighing between 150 and 200 g, water loss is about 50% evaporative (EWL) and 50% urinary–faecal. In smaller birds, such as Stark's Lark and the Grey-backed Finchlark, the EWL is about five times urinary–faecal water loss (Willoughby 1968).

Under moderate conditions the Zebra Finch with a body mass of a little more than 11 g loses about 2.1 g of water (about 20% of body mass) per day by evaporation, but produces only about 1 g of metabolic water per day, so it produces metabolically only about half of the water it loses (Bartholomew and Cade 1963). The higher the T_A the more water will be lost by evaporation (EWL). EWL increases fourfold between T_A 30-40 °C, while oxygen consumption increases only a little more than twofold. Reliance on metabolic water as the sole source of water, other than a small amount of preformed water, is probably rare or nonexistent in small birds.

Birds generally lose no more than 40-50% of their resting heat production by evaporation at high T_A (Cade 1964). An exception is the Poorwill *Phalaenoptilus nuttallii* whose basal metabolic heat production is only about a third of that of most birds of equivalent body mass, and which consequently can lose more heat by gular flutter than it produces metabolically. Some waxbills, including the Zebra Finch, can lose significant amounts of heat by panting.

Consumption of *ad libitum* water is usually inversely related to body mass; the Budgerigar and the Zebra Finch, however, drink less than predicted for their body mass. These two species can also maintain body mass in the absence of drinking water. Their water consumption usually increases with increasing T_A.

Neither metabolic rate nor the rate of EWL of Stark's Larks and Grey-backed Finchlarks appeared to be affected by water deprivation, but their mean rates of EWL were less than 9% of the body mass per day – less than half the rate expected when compared with those of other small passerines. There appears to be a definite adaptive reduction in pulmocutaneous water loss in these birds. Both species reduced their seed intake when deprived of water, that for the unwatered Grey-backed Finchlark being 81% and that for Stark's Lark being 60% of the mean for watered birds (Willoughby 1968). The water content of the whole seed was 8.9%, but the birds always husked the seed by cracking it before eating it, and the kernel had a water content of about 9.1%. The water content of the excreta (faeces and urine together) was high in watered birds, but in water-deprived individual birds it was as low as 30% in the Grey-backed Finchlark and 25% in Stark's Lark; the mean for both species was a little over 50% (Willoughby 1968). Other comparable species of birds had excretory water values of 68.1% (Budgerigar), 65.3% (Zebra Finch) and 57% (Black-throated Sparrow).

In the Namib, T_A is seldom in excess of 40 °C, so these larks may not often need to use evaporative cooling, further enabling them to survive with a minimum of free water. They also remain inactive in the shade during the heat of the day and may augment their water intake by eating insects such as ants, termites and beetles, as well as green vegetation when it is available (Willoughby 1968).

Evaporation of water in the Zebra Finch amounts to about 20% of the body weight per day, whereas cloacal loss of water is only about 13% at most (Dawson and Bartholomew 1968). The inverse relationship between evaporative water loss and body mass in birds is generally independent of systematic affinities and habitat: desert birds appear to be little better at controlling EWL than non-desert species of comparable size. Some exceptions to this include the Budgerigar, Bourke's Parrot *Neophema bourkii*, the Scaly-feathered Finch, the Zebra Finch and the Black-throated Sparrow. These birds can survive on a dry diet for unexpectedly long periods and must reduce their evaporative water loss; the Budgerigar does this in part by curtailing activity during water deprivation, but the other species appear not to do so.

Birds that have been studied lose more water by EWL (0.9-3.0 mg H_2O/ml of O_2) than mammals (0.50-0.59 mg H_2O/ml of O_2 in some desert rodents). This may be the result of birds expiring air at a higher temperature and therefore a greater relative humidity than do mammals. Also, their metabolic rates are higher, so that the rates of gaseous exchange are higher.

Evaporative water loss is greatly increased, by 2.5-5 times, during heat dissipation at high ambient temperatures (above 40 °C). Abert's Towhee *Pipilo aberti* frequently encounters such high ambient temperatures in the southwestern desert of the United States, often for periods exceeding 5 h during the day, during which time it must evaporate at least 10% of its body mass (46 g). Similarly the Zebra Finch evaporates about 30% of its body mass in a period of 5 h at an ambient temperature of 43.4 °C.

At moderate ambient temperatures (25 °C) oxidative water production is always less than EWL in all birds of a weight range from less than 10 g to about 150 g (Fig. 4, p. 368 in Dawson and Bartholomew 1968). In small birds (Costa's Hummingbird *Calypte costae* 3 g, and North American Cardinal *Cardinalis cardinalis* 40 g) the maximal oxidative water production per day does not even come close to offsetting the water lost by evaporation (Table 5.8).

Table 5.8 does not account for water lost in the faeces and urine, so the species shown lose even more water than the data show, though some species appear to be exceptions (e.g. Zebra Finch, Black-throated Sparrow and some others already mentioned). The disparity between oxidative production of water and EWL is inversely related to body mass, so that small birds are far more dependent on succulent foods and drinking water than are larger birds. Heat stress will aggravate this discrepancy because of the need for

Table 5.8. Oxidative production and evaporative loss of water at T_A 23-25 °C. (Dawson and Bartholomew 1968)

Species	(a) Body Mass (g)	(b) Maximum oxidative water production (g/day)	(c)Evaporative water loss (g/day)	c/a (%)	b/c (%)
Costa's Hummingbird *Calypte costae*	3	0.36	0.87	29	41.38
North American Cardinal *Richmondena cardinalis*	40	1.68	2.40	6	70.00

Table 5.9. Respiratory and cutaneous water loss (EWL) at T_A 30 °C and water-vapour pressure of 10 mm Hg. (Bartholomew 1972)

Species	Mass (g)	Evaporative water loss (EWL)		
		Respiratory	Cutaneous	
		mg/g/h	mg/g/h	Total EWL (%)
Zebra Finch *Poephila guttata*	12.5	3.3	5.6	63.0
Spotted-backed Weaver *Ploceus cucullatus*	42.6	3.3	3.4	50.8
Painted Quail *Coturnix chinensis*	42.3	2.7	2.2	46.8
Budgerigar *Melopsittacus undulatus*	31.6	3.7	5.3	58.9

Table 5.10. Body weight (g) and percentage of cutaneous evaporative water loss (%CEWL) of total evaporative water loss at T_A 30-35 °C for eight species of birds (Lasiewski et al. 1971).

Species	Mean body weight (g)	%CEWL
Zebra Finch	12.5	62.9
Budgerigar	31.6	58.9
Painted Quail	42.3	44.7
Spotted-backed Weaver	42.6	50.8
Poorwill	43.2	51.3
Feral Pigeon (Rock Dove)	299.9	74.0
Roadrunner	274.2	51.0
Ostrich	88 000.0	< 2.0[a]

[a]The result for the Ostrich is suspect, since it was extrapolated from CEWL through the skin of the legs only.

evaporative cooling. Even so, the Budgerigar and the Scaly-feathered Finch drink far less than might be expected for birds of their small body size. The Budgerigar is also unusual in not increasing its water intake significantly under hot, dry conditions in the way that Ground Doves *Columbina passerina*, Brown Towhees *Pipilo fuscus*, Abert's Towhees and House Finches do.

Cutaneous evaporative water loss (CEWL) has been shown to be significant in a number of species of birds, including the Feral Pigeon, Painted Quail *Coturnix chinensis*, Spotted-backed (Village) Weaver *Ploceus cucullatus*, Budgerigar and Zebra Finch (Bernstein 1971). Only the last two species are inhabitants of deserts. Simultaneous measurements of respiratory EWL (REWL) and CEWL were made. Cutaneous EWL is almost equal to REWL, being from 47-63% of the total EWL (Table 5.9).

This and other studies (Table 5.10) showed that CEWL varied from 2.1-5.6 mg H_2O/g per h and amounted to 44.7-62.9% of the total EWL in the four experimental species given in Table 5.9, excluding the Feral Pigeon. EWL must therefore play an important part in avian thermoregulation. At high T_A s of 40-42 °C CEWL accounted for about half the total evaporation in the Painted Quail, thereby dissipating much of the heat produced by the body.

The conclusion is that the range of CEWL in birds may be far greater than hitherto recognized or than observed by Bernstein (1971). Previous interpretations of CEWL on the basis, for example, of low CEWL through the legs of the Ostrich, extrapolated to the skin of the rest of the bird, suggest that CEWL would account for only about 2% (Schmidt-Nielsen et al. 1969), whereas in the Feral Pigeon it accounts for as much as 74% of the total EWL.

Because birds lack sweat glands, it has been assumed that CEWL in birds is negligible. Studies on EWL in the Roadrunner and the Poorwill showed that the CEWL accounted for a little more than half the EWL (Lasiewski et al. 1971; Table 5.10), as in the case of the four species studied by Bernstein (1971). There is no clear correlation between the amount of CEWL and habitat in the sample in Table 5.10. In the Roadrunner and the Poorwill CEWL accounts for the loss of 14-20% of metabolic heat production at T_A 30-35 °C.

A source of water conservation in some birds, which has been mentioned in the Introduction, and to which little attention has been paid, is the countercurrent effect produced in the nasal passages by inhaled and exhaled air. Exhaled air is saturated with water vapour; the higher its temperature, the more water it carries and the greater the degree of EWL (Bartholomew 1972). At T_A below T_B inhaled air cools the respiratory tract, resulting in subsequent cooling of exhaled air and some condensation of its water vapour. The lower the temperature of the respiratory tract, the greater the amount of water vapour recovered. In the Cactus Wren *Campylorhynchus brunneicapillus* at T_A below 35 °C the temperature of exhaled air is always less than that of inhaled air, so that countercurrent cooling must be important in water conservation. Since this mechanism operates in all birds, even up to the size of the Ostrich (Withers et al. 1981; see Sect. 5.6.1), in all kinds of habitats, it is a useful preadaptation to life in the arid zone, and not a new feature of desert birds in particular.

5.5 Arid-Adapted vs Non–Arid–Adapted Birds

In terms of the osmoregulatory and excretory functions mentioned above, the question arises as to whether birds occupying arid zones indeed have some physiologically adaptive advantage over those of other habitats. The relative physiological performances of arid-adapted birds and those adapted to mesic habitats have been demonstrated in four selected studies.

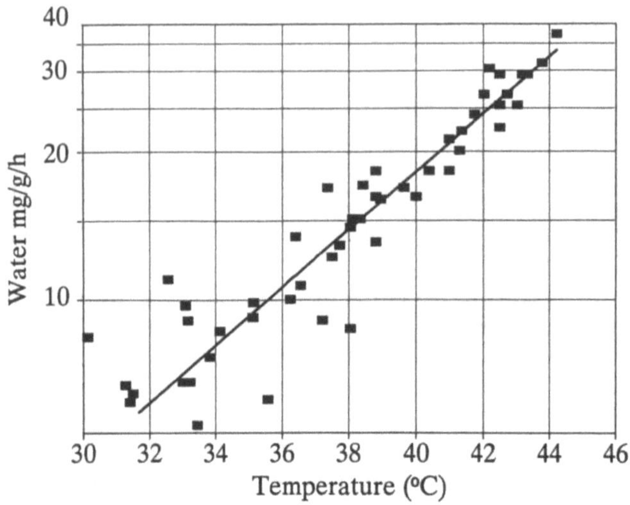

Fig. 5.9. Log of evaporative water loss in Abert's Towhee *Pipilo aberti* (expressed as mg water lost/g body weight/h), relative to ambient temperature. (Redrawn from Dawson 1954)

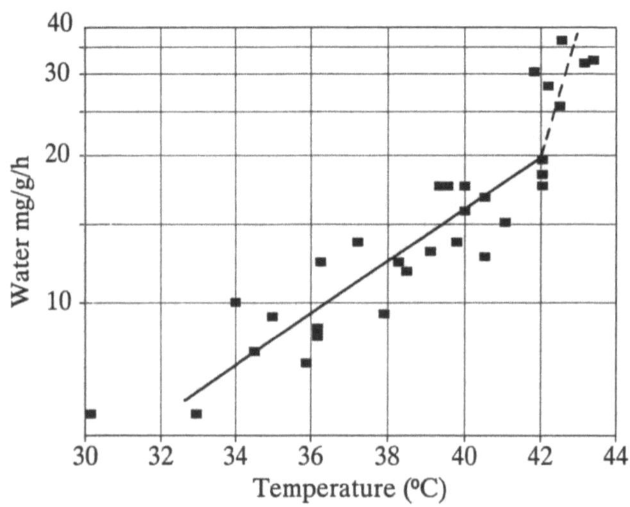

Fig. 5.10. Log of evaporative water loss in the Brown Towhee *Pipilo fuscus* (expressed as mg water lost/g body weight/h), relative to ambient temperature. (Redrawn from Dawson 1954)

The first is that of Dawson (1954) on two species of towhee (Nearctic fringillids). Abert's Towhee is an inhabitant of arid North America where very high summer temperatures prevail. The Brown Towhee is widespread in more mesic habitats, but does penetrate to some desert areas whose summer temperature regimes are less severe than those experienced by Abert's Towhee. At high T_A (39 °C) Brown Towhees drank significantly more water than Abert's Towhees and the Browns succumbed to water deprivation more easily than the Abert's Towhees. Brown Towhees at T_A 39.5 °C and RH 36% lost about 35% of their normal body weight/day by evaporation. An Abert's Towhee at T_A 39.5 °C and RH 45% lost at least 39% of its normal body weight/day by evaporation. The water loss of Abert's Towhee was surprisingly greater than that of the Brown Towhee. However at T_A above 40 °C EWL in the Brown Towhee increased more rapidly than that of Abert's Towhee (Figs 5.9 and 5.10).

At T_A 23 °C the drinking rates of the two species were not significantly different, but at T_A 39.5 °C Abert's Towhee drank less (only 1.9 times more than at 23 °C) than did the Brown Towhee (3.6 times more than at 23 °C). Comparison of the daily EWL with daily water consumption of the two species showed that Abert's Towhee drank 5% more than it lost by EWL, while the Brown Towhee drank 28% more. The arid-adapted Abert's Towhee is clearly the better performer under heat and water stress.

The second such study is that of MacMillen and Trost (1966) on two of the most abundant birds in the Sonoran Desert in summer, the White-winged Dove (Fig. 5.2) and the Inca Dove of the southeastern United States. Without prior water deprivation, White-winged Doves maintained body mass on an *ad libitum* diet of dry seeds (water content 9-10%) and tap water and 12.5% seawater, but lost mass on higher concentrations than this. When deprived of water before experimentation they maintained mass on 25% seawater and actually gained mass on concentrations of 37.5% seawater. Both experimental groups lost mass on 50% seawater. Inca Doves drank water of any concentration without being deprived of water beforehand, merely increasing their water intake with increasing concentration, but they lost mass at concentrations of 37.5% seawater and higher and when deprived of water altogether.

The drinking solution of 37.5% seawater has an osmolarity of 354.6 mOsmol/l; the plasma osmolarity of White-winged Doves on solutions up to this concentration varied from a mean of 335.4 mOsmol/l to a mean of 353.2 mOsmol/l (therefore hyposmotic to the drinking solution). On 50% seawater White-winged Doves lost mass and had significantly elevated plasma osmolarities (mean of 428.4 mOsmol/l) indicative of a state of partial dehydration.

Results for Inca Doves were more variable, possibly because of the need to dilute their blood samples to get enough solution for testing, but they appeared to have similar plasma osmolarities to those of White-winged Doves at seawater concentrations of up to 37.5%. Generally, however, Inca Doves could not tolerate salinity greater than 25% seawater.

When deprived of water altogether, both species lost mass linearly, but the larger White-winged Doves lost it more slowly and tolerated greater loss of mass than did the smaller Inca Doves. The White-winged Doves were also better at rehydration than Inca Doves which had lost the same percentage of body mass. Differences in tolerance may be related to the absolute abundance of nephrons in the kidneys, those of the larger birds being more numerous and therefore better able to excrete excess salt. "Thus body size dictates urine-concentrating capacity of the bird and hence its salinity tolerances" (MacMillen and Trost 1966).

These arid-zone doves, especially the White-winged Dove, could maintain body mass on a diet of seeds augmented with succulent foods, such as tomatoes (95% water by mass), Saguaro *Cereus giganteus* (85% water) and Prickly Pear *Opuntia occidentalis* (85% water).

Neither species of dove shows any special economy of water utilization associated with an arid environment. Differences in water relations were related particularly to body size, the larger birds being better at tolerating water loss and saline waters, as well as at rehydrating after water deprivation. The intermediate-sized American Mourning Dove was intermediate in its performance between the White-winged and Inca Doves. A further difference between White-winged and Inca Doves is that the former are strong fliers and range throughout the desert, drinking greedily only twice a day (as do American Mourning Doves) at a variety of water sources, whereas Inca Doves occur only in and around human settlements and drink small amounts of nearly fresh water frequently throughout the day.

The third comparative study is that of Cade et al. (1965) between two species of estrildids: Zebra Finches inhabit arid conditions in Australia, whereas the Black-rumped Waxbill *Estrilda troglodytes* lives in riverine and moist grassy places in Africa. They represent extremes of habitat utilization among estrildids. Under conditions of moderate T_A (20-22 °C) and RH (20-40%) and on a diet of air-dried seeds (about 10% water by mass), water-deprived Black-rumped Waxbills died within 3 days, whereas Zebra Finches survived indefinitely, some for as long as 250 days without water. The waxbills follow the general pattern for small passerines, whereas the Zebra Finch shows remarkable ability to withstand water deprivation. With *ad libitum* water, the waxbills drank on average 42% of their body mass per day, whereas the Zebra Finches drank only 24% of body mass per day.

With an increase in T_A to 30-40 °C the waxbills showed a great increase in water consumption to over 100% of body mass per day, but the response of Zebra Finches was less clearly defined; they did increase their water intake, but seldom to as much as 100% of body mass per day. At high T_A (from about 39 °C) the waxbills panted almost continuously, but the Zebra Finches did not start to pant until T_A reached 42 °C. At rest below these upper critical temperatures (T_{UC}), the Black-rumped Waxbills lost about 29% of their body mass per day by evaporation, whereas Zebra Finches lost only about 20.5%. In the absence of water to drink the waxbills could not produce enough metabolic water to offset EWL, whereas Zebra Finches could.

The waxbills could not maintain body mass when drinking NaCl solutions of more than 0.15 M and some individuals increased their daily water consumption to 200-400%. Zebra Finches, surprisingly, also had a limited tolerance to solutions of NaCl: some captive individuals died after only 6 days on 0.1 M NaCl, but birds conditioned to high concentrations could maintain mass on solutions as high as 0.4 M NaCl. Water-deprived Zebra Finches given 0.4-M solutions died if they drank too much, but survived if they reduced their water consumption down to almost nothing. These experiments seem to indicate that birds on a restricted water intake had enhanced production of antidiuretic hormone (ADH) and could therefore withstand further deprivation and/or highly saline solutions in small quantities. Both estrildid species could distinguish between distilled water and solutions as dilute as 0.05 M NaCl, but not as dilute as 0.03 M.

All in all, the Zebra Finch is considerably better adapted to arid conditions than the Black-rumped Waxbill by having a greater tolerance to heat and a better water economy.

A fourth comparative study, by Skadhauge (1972), involved the Domestic Fowl and the Budgerigar. I am not sure on what grounds the Domestic Fowl counts as a xerophilic bird, as indicated by the title of this paper, but the information provides a useful comparison between the originally forest-adapted Domestic Fowl and the unequivocally arid-adapted Budgerigar. Glomerular filtration rate (GFR) is usually around 2-3 ml/kg for birds in general; from the hydrated to the dehydrated state this dropped by about 23% in the Domestic Fowl and by about 37% in the Budgerigar. In the dehydrated state the fractional water absorption in the tubules was at least 99% of filtered water; this is associated with the limited concentrating ability of the avian kidney. The presence of a caecum in the Domestic Fowl is associated with water resorption; if the caeca are extirpated, the urine is more dilute. The Budgerigar, however, lacks caeca.

Maximum urine osmolarity in the Domestic Fowl was around 600 mOsmol/l and in the Budgerigar around 1100 mOsmol/l. Arginine–vasotocin (AVT) induced antidiuresis in an action similar to that observed during dehydration, so that reduced GFR during dehydration may be the effect of this hormone. In the dehydrated Budgerigar the preformed water in the food (dry seed) and the water of oxidation (metabolic water) just balanced the evaporative and ureteral urine water loss; the Budgerigar can remain in water balance only if cloacal resorption or ureteral urine equals the water loss in the faeces.

5.6 Some Case Studies

With that background of avian physiological performance, it is helpful and interesting to look at some case studies on selected arid-zone birds, from which a set of conclusions may be drawn concerning their water-related adaptations to their apparently stressful habitat.

5.6.1 Ostrich *Struthio camelus* (Fig. 3.6)

With increasing dehydration the volume of urine and the frequency of urination decreased in the Ostrich (Louw et al. 1969). The consistency of the urine also changed from clear and watery to highly viscous in a brown mucus. The lining of the ureter is found to be rich in goblet cells, so that at least part of the mucus originates in the ureters. The use of mucus as a lubricant for the uric acid crystals is a significant factor in water conservation, and may allow the passage of more concentrated urine than was suggested possible by Schmidt-Nielsen et al. (1963).

When the birds were hydrated, the Na^+ and K^+ concentration fell sharply (Louw et al. 1969); during dehydration, however, concentration of both these electrolytes rose initially, but that of Na^+ dropped and remained low, suggesting the action of aldosterone which promotes Na^+ retention. As the osmoconcentration of urine increased during dehydration to a maximum of about 780 mOsmol/l, that of the plasma remained static at around 300 mOsmol/l, so that the maximum urine osmoconcentration was only 2.7 times that of the plasma (nowhere near as high as that in some arid-adapted mammals). The function of the nasal glands in osmoregulation was not tested.

The water content of the faeces also decreased during dehydration by a mean of 18% (Louw et al. 1969). Cloudsley-Thompson and Mohamed (1967) showed that the Ostrich can maintain body mass on a regime of 20% seawater and can withstand a 25% loss of mass during dehydration.

It has always been assumed the expired air in vertebrates must be saturated with water vapour, regardless of temperature (Withers et al. 1981). Although the assumption holds good for *Homo sapiens*, it does not for the Ostrich. Adult Ostriches were studied at T_A 28-32 °C and RH 30-36% by using deep sensors in the large nasal passages. The T_B of the experimental birds varied from 39-41 °C. The temperature of the expired air was 3-4 °C lower than T_B. Inspired air had an RH corresponding to that of ambient air, whereas air expired through the nostrils had an RH of 75-100%; air expired through the mouth had an RH of 100%. The nasal chamber of the Ostrich is divided into a lower and an upper portion by an incomplete septum: the RH of expired air was significantly lower in the upper chamber than in the lower. The mechanism whereby the RH of expired air is reduced is not known, but it results in a saving of about 50% of the water added by the bird to the inspired air. The saving to the Ostrich of expiring unsaturated air is considerable and may be of critical value to the bird in an arid environment when drinking water is in short supply.

5.6.2 Sand Partridge *Ammoperdix heyi* and Chukar Partridge *Alectoris chukar*

The Sand Partridge and the Chukar Partridge occur sympatrically over much of the Negev Desert, but the Sand Partridge can live in much more arid places where Chukars are absent. Yet experimental evidence shows that neither species is at any greater advantage or disadvantage than the other with regard to water balance, so that no explanation can be offered as to why the Sand Partridge seems to better able to inhabit the more arid regions (Pinshow et al. 1983). Birds kept on a different dietary regime in captivity might show greater physiological differences in their water balance.

Neither the Sand Partridge nor the Chukar Partridge was short of drinking water in the wild in the Negev Desert (Thomas et al. 1984). The compositions of their fluids (plasma, urine, rectal luminal and voided fluid) were intermediate between those of watered and water-deprived captive birds of the same species. Indications are that the lower intestine modifies the post-renal urine only slightly before excretion.

As Chukar Partridges shifted from a diet of dry food in summer (mainly seeds; water content 57%) to one of succulent food in winter (green vegetation; water content 82%), their water-turnover rate increased from 101 to 266 ml/kg per day, but it can reach as high as 390 ml/kg per day in a season of heavy rains (Alkon et al. 1985). Seasonal changes in the rate of water turnover are linked therefore to diet and are a function of the intake of preformed water in the food. Greens are scarce in the wild in summer, but captive Chukars eat greens readily at any time. Chukars cannot maintain their body mass on a diet of greens only, however, possibly because of a digestive bottleneck; they must have seeds as well, at least initially, since the gut morphology takes about 25 days to change to accommodate a diet of greens. This study is rather weak, I think, and contains more speculation than information.

5.6.3 Gambel's Quail *Callipepla gambelii*

Gullion (1960) showed that Gambel's Quail in Nevada, USA, needs no drinking water if it inhabits desert valleys with Desert Riparian Shrub or Thorn Shrub, or where there are irrigated crops; a combination of succulent plants and high humidity provides enough water even at the hottest times of the year, so the birds need not drink at all. On the desert uplands, where temperatures are more moderate, the quails' water needs may be met by water taken in with succulent plants and insects. Some metabolic water may augment this water supply, especially where the seeds of certain legumes (*Lotus* and *Astragalus*) are important sources of food. Rain brings out succulent seedlings of desert ephemerals (especially *Erodium*), which are an important source of water.

The critical period for quails in the uplands is mid-June to mid-July when mean daily temperatures are around 38 °C or higher, the succulent spring annuals have withered, and the summer thunderstorms have not yet begun. A useful game-management technique at this time of the year is the provision of artificial watering points.

In this species, as in other North American quails of about the same body weight (150-200 g), water loss is about 50% EWL and 50% urinary–faecal (Bartholomew 1972). When deprived of water, Gambel's Quails do not reduce their glomerular filtration rate (GFR) more than do other birds from more mesic habitats. A reduced GFR is a primary renal mechanism for the conservation of water under such conditions (Williams et al. 1991). Plasma osmolality of normally hydrated quails varied from 326-389 mOsmol/kg of water, averaging 358 mOsmol/kg H_2O. After 4-5 days without water, but with ad libitum food, plasma osmolality increased by 10% from a mean of 358 mOsmol/kg H_2O to 393 mOsmol/kg H_2O, but it had a smaller range of 380-413 mOsmol/kg H_2O than in normally hydrated birds.

In normally hydrated birds GFR ranged from 10.6-27.4 ml/h (mean 17.7 ml/h); this decreased in water-deprived birds by 58.2% to a mean GFR of only 7.4 ml/h. The concentration of plasma Na^+ is much greater in normally hydrated birds than in water-deprived birds, indicating a much tighter control of this electrolyte when under water stress. GFR in Gambel's Quail is closely related to water scarcity in the environment, falling markedly in response to water deprivation. Similarly, as plasma osmolality increases, so GFR declines. How this is achieved is not altogether clear, but it seems as if an increase in extracellular osmolality stimulates the release of arginine vasotocin (AVT) from the hypothalamus, which causes a constriction of afferent arterioles supplying the glomeruli of the RT (reptile-type) nephrons in the kidneys. Why the MT (mammal-type) nephrons are unaffected is also not clear.

In a study by Williams et al. (1991), water-deprived quails had a $U{:}P_{osm}$ averaging 1.6 (maximum = 2). However, it has been suggested (Skadhauge 1974) that birds may reflux urine from the cloaca into the lower intestine where further resorption of water and electrolytes may occur, in the same way that the distal nephron does in mammals. Such refluxion may not occur, however, when the birds are dehydrated (Thomas et al. 1984). Such water resorption is solute-linked since it follows that transport of Na^+ across the epithelial tissues is accompanied by a passive flow of water in the same direction. In the study by Williams et al. (1991), the quails had a plasma concentration of 393 mOsmol/kg H_2O and a urine concentration of 637 mOsmol/kg H_2O, a difference of 244 mOsmol/kg H_2O. Under such conditions, the birds would lose water rather than gain it, if they refluxed it. The plasma osmolality of birds can vary by as much as 30%, while that of mammals seldom varies by more than about 2%.

The predicted allometric GFR for normally hydrated Gambel's Quail weighing 158 g would be 40.3 ml/h, a value three times greater than that measured (Williams et al. 1991). Thus this species seems to have an unusually low GFR, even when hydrated, which may reflect its relatively small kidney mass, which is about 30% less than predicted from body mass. The low GFR may also be associated with the birds' unusually low field metabolic rate (FMR) which is only 40% of that predicted for its body mass: a reduced FMR results in smaller REWL (because of lower oxygen demands) as well as the production of fewer metabolic waste products and therefore a reduced GFR and attendant water loss in the urine flow.

5.6.4 Sandgrouse *Pterocles* spp. (Figs. 3.8 and 3.9)

Sandgrouse in the deserts of Morocco live in thermally stressful habitats (T_A up to 55 °C), are diurnal and nest in the open (Thomas and Robin 1977). They minimize heat loads by confining their most vigorous activity to the cooler parts of the day, remaining inactive in the heat and increasing their thermal insulation when T_A exceeds T_B. They conserve water by delaying evaporative cooling (by gular flutter) until T_A reaches 50 °C. They may also have a water-resorptive system in the kidneys and colon and they need to conserve salt, since their watering places appeared to be salt-deficient.

In the arid regions of Namibia, Namaqua Sandgrouse *Pterocles namaqua* and Double-banded Sandgrouse *P. bicinctus* drank little at T_A below 37.5 °C, but increased drinking significantly at higher T_A (Thomas and Maclean 1981). Dehydrated Namaqua Sandgrouse showed no signs of distress after three days and then drank only about 8 g H_2O/100 g of body mass; Double-banded Sandgrouse drank much more (12 g H_2O/100 g of body mass). Both species took a mean of 2.34 g of water per mouthful (= draught). Namaqua Sandgrouse drank 8-10 draughts per undisturbed drink in the field, which amounts to 19-23 g of water, or 11-14 g H_2O/100 g of body mass.

Water turnover for both species of sandgrouse was only 36-41% of the expected value on *ad libitum* water intake compared with that of other species under similar conditions, but it was 142-164% of the expected value of the minimum intake to maintain body mass. The low water-turnover rate reflects the low resting metabolic rate (RMR) and must contribute to the birds' ability to survive 3 days of dehydration in captivity. The conclusion from this research is that Namaqua Sandgrouse, at least, do not drink daily, but perhaps every 3-5 days. (Males with young may drink daily, however.) Drinking less often than every day avoids undue exposure to predators at the watering places.

Water is further conserved by heat storage, since the birds are not cooling by evaporation whenever T_B reaches a certain level. The potential water saving (W in g/bird) by heat storage can be calculated by the equation:

$$W = c' \, m' \, (\Delta T_B)/L$$

where c = specific heat taken as 3.3 J/g_{bm} per °C, m = body mass in grams, ΔT_B = change in body temperature and L = latent heat of vaporization taken as 2400 J/g water at 40 °C. For hydrated Namaqua Sandgrouse the value of W would be 0.2 g H_2O/100 g of body mass, or about 2.4% of the estimated daily water turnover with free access to water. When dehydrated the saving would be even greater, around 7.4% of the observed daily loss of body mass during dehydration. Because of their greater body size and more labile T_B, Double-banded Sandgrouse would be able to save even more than Namaqua Sandgrouse.

Fresh sandgrouse faeces are very dry and it is difficult to obtain fluid from the lumen of the hindgut. Maximum osmolarities of 500 mOsmol/l were measured in the rectal fluid; serum osmolarity was 300 mOsmol/l. Between

36 and 53% of the rectal-fluid osmolytes was ascribable to Na^+; the composition of this fluid indicated effective resorption of salts and water and was therefore compatible with the dry appearance of the excreta.

5.6.5 Rock Dove (Feral Pigeon) *Columba livia*

Rock Doves (mass about 220 g) lose up to 16% of their body mass after more than 48 h of heat exposure (T_A up to 50 °C) in the absence of drinking water at RH < 20% (Arad et al. 1987). The particular group of seven experimental pigeons used in this study survived T_As of up to 60 °C without undue distress as long as they had access to water and were normally hydrated, maintaining a stable T_B while losing water cutaneously by evaporation in the absence of panting or gular flutter; they also bred normally under these conditions. When given water to drink, the birds regained their original body mass within 30 min. Cutaneous evaporation was very low at a thermoneutral temperature of 30 °C, whether the birds were hydrated or dehydrated, but it increased about sixfold during heat exposure in normally hydrated birds. Even so, these hydrated birds neither panted nor gular fluttered, but when dehydrated six of the seven experimental birds stopped cutaneous evaporation and began to pant. Dehydrated birds also become significantly hyperthermic at high T_As.

Although cutaneous and respiratory water losses were not measured simultaneously, it was possible to make a rough estimate of their relationship to one another: in normally hydrated birds in the TNZ (thermoneutral zone), the ratio between the two pathways of water loss is about 1:1, but above the TNZ, CEWL accounted for nearly all water evaporated from the body. In contrast, dehydrated birds lost about 70% of their evaporated water via the respiratory route. Two of the benefits of this form of evaporative cooling are (1) that it avoids the risk of CO_2-washout from the blood via the lungs and consequent deterioration of the acid-base status; and (2) that it does not result in further heat production by the muscular activity of enhanced breathing movements during panting. No explanation has been advanced for the mechanisms of regulation of CEWL.

The RMR of the pigeons was 25-30% lower than expected for a 220-g non-passerine, possibly as an adaptation to desert environments where the T_A can be very high. This has also been shown for other heat-adapted columbiforms (e.g. Frith 1982; Withers and Williams 1990; Schleucher et al. 1991).

5.6.6 American Mourning Dove *Zenaidura macroura* (Fig. 4.5)

The normal plasma osmolarity of 12 experimental birds on *ad libitum* water was 372 mOsmol/l. When deprived of water this went up to 411 mOsmol/l (Smyth and Bartholomew 1966b). The Na^+ increased slightly though significantly, but the Cl^- concentration did not increase significantly. The birds could drink NaCl solutions of 0.1 M without ill effect, but could not tolerate 0.2 M NaCl. The higher concentration caused the plasma Cl^- to increase but not that in the urine. The Na^+ concentration in the plasma was unchanged. The maximum concentration of the urine, even when NaCl was administered by stomach tube, was 363 mEq/l (mean maximum for all birds = 327 mEq/l). The physiological performance of the American Mourning Dove is therefore no better than that of other non-desert birds tested.

5.6.7 Ground Dove *Columbina passerina*

The Ground Dove inhabits an arid climate in California where ambient temperatures are high (Willoughby 1966). It seldom lives far from water. If deprived of water, the doves lost mass steadily; some stopped eating and lost mass more rapidly. Loss of mass also occurred when the birds were put onto a regimen of 0.15 M NaCl, which must be close to their limit of tolerance, since they can hold their mass well on solutions of only 0.10 M NaCl. The minimum water requirement when drinking distilled water was about 10% (8.8-10.3%) of body mass per day; the mean *ad libitum* consumption of distilled water was 11.4% per day. Ground Doves appear to show no special adaptations to a hot, dry environment.

5.6.8 Plumed (Spinifex) Pigeon *Geophaps plumifera*

The Plumed (or Spinifex) Pigeon inhabits semi-arid and arid regions of Australia often, but not necessarily, with rocky and stony substrates. It flies relatively weakly and cannot therefore travel very long distances to water; thus it is found near permanent water and drinks intermittently (Withers and Williams 1990). It occurs in flocks of up to about 50 birds, usually travelling as compact groups from up to 2 km or more to watering places. The water is approached on foot, less often by flying in, and the birds drink and move away in a straggling line. They drink throughout the day, usually from 10:00 or 11:00 h. They are absent from otherwise suitable areas, if water is not available (Frith 1982). Their food is the seeds of grasses, especially spinifex *Triodia* where this occurs. They experience considerable water and heat stress in this environment, which includes some of the hottest places on earth. A study of their osmoregulatory physiology would be of the greatest interest.

5.6.9 Budgerigar *Melopsittacus undulatus*

Domesticated Budgerigars drank only 5% of their body mass per day under moderate T_A (24-28 °C), and at higher T_A increased their consumption to only about 10% of body mass; some birds did not even increase their water consumption over a T_A range of 20-30 °C (Cade and Dybas 1962). Individual variation allowed a partition of the birds into "drinkers" and "nondrinkers";

when given saline water (0.2 or 0.3 M NaCl) the drinkers increased their consumption, the non-drinkers decreased it; drinkers lost mass on salt solutions, non-drinkers held their mass. Most of the birds decreased food intake at higher T_A, whether drinkers or non-drinkers. When deprived of water, most of the birds maintained a uniform food consumption, but one bird increased its food intake considerably. Normally hydrated Budgerigars produced excreta with 75-80% water content. Dehydrated birds averaged only 60% of water in their faeces. Resting birds had an EWL of 1-2 g daily, but when active this went up to 2-3.5 g daily.

A Budgerigar weighing 30 g and consuming 4 g seed/day, of which it metabolizes about 90%, will produce only about 1.844 g water. Of this water, 0.65 g are lost in the excreta, leaving only 1.19 g for respiratory loss. On the basis of these calculations, a Budgerigar would hardly be able to maintain itself in water balance on 4 g of dry seed alone, yet it does so for long periods. The conclusion is that the experimental measurements of pulmocutaneous EWL are too high. Stress in captivity may have contributed to a greater EWL, and the T_A in the experiment was about 5 °C lower than the lower critical temperature (T_{LC}) of this species. Even so, water balance may be attained by a nondrinking Budgerigar only if it is as inactive as possible, especially at higher T_As.

Budgerigars also have a lower basal metabolic rate (BMR) than that predicted for birds of their size. Under natural conditions they become completely inactive during the midday hours. From the data it is likely that wild Budgerigars cannot survive for long without drinking. Heatwaves in inland Australia may result in the deaths of millions of Budgerigars.

5.6.10 Roadrunner *Geococcyx californianus* (Fig. 5.6)

The Roadrunner inhabits the most arid regions of North America, eats mostly animal food and drinks water under natural conditions, although it seems not to need it (Ohmart et al. 1970). When fed only white mice in captivity, in the absence of drinking water and at T_A of 30-32 °C, Roadrunners turned over a mean of 26.2 ml/day of water; when given water *ad libitum* at low T_A the mean water turnover was 41.8 ml/day. The difference was highly significant (Student's t-test; $P < 0.01$). When deprived of drinking water, the faeces showed a marked reduction of water content and much salt encrusted the nares. Mouse tissue contains about 67% water, which amounts to about 19 ml/day; the deficit between this intake and the 26 ml/day water turnover was presumably supplied by metabolic water derived from digestion of the mouse tissue. These tests were done under moderate conditions and show only a general capacity of birds to reduce water loss when deprived of drinking water. I do not believe that any great claims can be made for desert adaptations in the Roadrunner on this basis.

5.6.11 White-Throated Silverbill *Lonchura malabarica*

The White-throated Silverbill inhabits arid regions of North Africa, Arabia and southern Asia (Willoughby 1969b). Like some larks in the Namib Desert, the Silverbill can survive on a diet of air-dried seeds and no drinking water at moderate T_A and RH. Pulmocutaneous evaporation rates of birds deprived of water were significantly lower than those of birds allowed to drink *ad libitum*. The rate of EWL was only 12% of body mass per day in water-deprived birds - much lower than predicted for a bird with a mass of only 9-12 g. The Silverbill has an upper critical temperature of between 42 and 44 °C, which is high for a small bird and probably, together with its low rate of EWL, an adaptation to a hot, arid environment.

5.6.12 Zebra Finch *Poephila guttata*

Under experimental conditions of water deprivation (0.25 ml/48 h or 2.1% of body weight) Zebra Finches showed a slight loss of weight, but most maintained body weight at > 95% of normal. In the total absence of drinking water one bird gained 0.8 g, another 0.2 g and a third stabilized at 0.8 g below normal (Calder 1964).

Mean food intake was 3.04 g/day in water-restricted birds and 2.9 g in birds with ad libitum water (no significant difference). The preformed water content of air-dried millet seed was 7.7%, so the preformed water intake was 0.24 ml/day for water-restricted birds.

Droppings from birds on *ad libitum* water typically had a solid core surrounded by fluid and a mean water content of 80.2%; those of water-restricted birds (0.25 ml/48 h) were solid, adhered only slightly to paper and had a mean water content of 65.3%. The mean daily cloacal water loss of birds on ad libitum water was 1.28 g, while that of water-restricted birds dropped to 0.54 g.

Adaptations of the Zebra Finch to extremes of heat and aridity include:

1. Reproductive activity is timed to coincide with favourable conditions after rain.
2. It can survive with very little drinking water - as little as 1 ml/week - or highly saline water (0.6 M).
3. Water is drunk by sucking ("aspiration") to use every available droplet.

Calder (1964) showed that motor activity was minimized when the birds were heat-stressed and that they tolerated limited hyperthermia rather than use evaporative methods of heat loss: T_B rose from 41 °C in the TNZ to a maximum of 43-43.5 °C above TNZ, at which point evaporative cooling was resorted to, but without the advantages of reduced RMR and gular flutter of extensive oral surfaces as in the Poorwill (Bartholomew et al. 1962).

Under moderate climatic conditions Zebra Finches can survive without drinking water for up to 18 months, but bill growth becomes irregular and the plumage loses condition (Sossinka 1972). There is some individual variation in the capacity to withstand water deprivation. Only 52% of

21 birds survived 18 months, 76% for 6 months and 90% for 3 months. There was an initial rapid loss of body mass which soon became compensated for. It is suggested that osmoregulatory adaptations to the arid environment are older phylogenetically than adaptations involving the regulation of the timing of breeding.

In order to keep cool, a Zebra Finch exposed for 5 h to T_A 43 °C loses water by evaporation amounting to about 30% of its body mass (Wiens 1991). Water lost in excreta may be less in desert birds than in birds of mesic habitats and they can reduce water lost in this way when dehydrated: Zebra Finches and Budgerigars produce excreta with a water content of 80% when on a diet of unrestricted water; when dehydrated they reduce this to 55-60%. Larks and Black-throated Sparrows can produce faeces with as little as 30% water. Much of the reduction in water loss during dehydration, however, is the result of restricted cutaneous evaporation (Lee and Schmidt-Nielsen 1971). Under moderate T_A in the laboratory some small granivorous birds (< 30 g body mass) can maintain water balance without drinking free water, by reducing evaporative, faecal and cloacal water loss. Part of their ability to do so seems to lie in their higher lower critical temperature (T_{LC}); since the rate of EWL decreases with decreasing T_A below the T_{LC}, and the production of metabolic water increases, there is a T_A at which water loss is balanced by metabolic water production. The bird is independent of water below this T_A. By increasing T_{LC} the value of T_A at which this balance is achieved is increased. There is a cost involved, however, in that the bird must expend more energy, but the benefit accruing from higher metabolic water production must offset this cost to some extent. This is probably especially important in granivores. Few desert birds are obligate granivores; those that are (such as doves and sandgrouse) must drink water. Nevertheless granivory is the "prevalent dietary mode" in desert birds generally (Wiens 1991). In the Namib and Kalahari, only 25-30% of species are granivorous, but they are the most numerically abundant species, as well as being gregarious (Maclean 1974c).

5.6.13 Sage Sparrow *Amphispiza belli*

The Sage Sparrow is an inhabitant of the more northerly deserts of North America (Moldenhauer and Wiens 1970). It is largely granivorous and occurs far from surface water. Males weigh 16.4-20.2 g with a mean of 18.4 g ($n = 26$) and females 16.2-19.8 g with a mean of 17.3 g ($n = 7$). The differences between the weights of the sexes are not significant at the level of $p = 0.01$ (t-test).

Water consumption of captive birds on a dry diet was sometimes as high as 100% of the body weight per day, but this was reduced when they were provided with succulent food [lettuce and mealworm (*Tenebrio*) larvae]. The birds could not maintain their body weight on a dry diet when deprived of water, losing 5.2% of their initial weight per day, and could not survive more than about eight days without water. They also reduced their intake of food, but increased their motor activity (searching for water?). However, the birds could maintain their body weight in the absence of drinking water when provided with succulent food; the mean intake of preformed water was about 2.5 ml/bird per day.

Sage Sparrows on *ad libitum* water lost about 4.9 g/day of water in the urine. When drinking water was reduced to 2.0 ml/day of water, the excretory water loss was reduced to 1.8 g/day by reducing the water content of the faeces. Birds maintained on mealworms and mash had a daily excretory water loss similar to that of birds on restricted water intake. Sage Sparrows were not very good at utilizing saline water: they could not tolerate solutions of NaCl of 0.25 M and above.

In the wild Sage Sparrows satisfy their summer water requirements by eating succulent food and insects, since they breed many kilometres from the nearest drinking water. Unlike the congeneric Black-throated Sparrow, which can survive complete deprivation of significant amounts of preformed water, Sage Sparrows lack any unique physiological adaptation to a desert environment.

5.6.14 Black-Throated Sparrow *Amphispiza bilineata* and Rock Wren *Salpinctes obsoletus*

Black-throated Sparrows and Rock Wrens are inhabitants of the desert regions of western North America (Smyth and Bartholomew 1966a). The sparrows feed mainly on seeds and appear to be more arid-adapted than any other North American seedeater. They commonly occur far from water on barren mountain slopes. Rock Wrens on the other hand are insectivorous and not confined to desert regions.

Black-throated Sparrows meet their water needs in winter and spring by feeding on green vegetation (including grass) and insects. At other times of the year when the food is dry seed, they drink water, mostly at mid-morning and mid-afternoon, but throughout the daylight hours. They also forage actively at T_A above 40 °C, but usually seek shade, pant and hold their wings away from the body. In captivity the Black-throated Sparrow drinks about 30% of its body mass per day (tap water). It also drinks water up to 0.2 M NaCl, but decreases its water intake at concentrations higher than this, maintaining body mass at concentrations up to 0.4 M. From 0.5 M upwards the birds lose mass. They can survive on a diet of dry seed (water content 9%) without drinking at all and without reducing their activity.

When fed NaCl by stomach tube, Black-throated Sparrows excreted Cl^- at concentrations greater than 350 mEq/l. Some individuals excreted 500 mEq/l and one even 703 mEq/l. Watered birds had a water content of 80% in their non-liquid excreta, while water-deprived birds reduced this to 57%.

Rock Wrens never drink, even when foraging near water. The Cl^- concentration of the birds' urine varied between 50 and 167 mEq/l under normal conditions, but could be raised to a maximum of 387 mEq/l when artificially fed solutions of NaCl. The Rock Wren shows no special adaptations to its arid environment, and can escape into rock crevices in the heat of the day.

5.7 Conclusions

5.7.1 Drinking Saline Solutions

Arid-adapted birds may drink NaCl concentrations as high as 50% seawater (0.3 M), but prefer solutions of no greater than 25% seawater. The Rock Parrot of Australia is exceptional in being able to maintain body weight when drinking 100% seawater (0.5 M). Some subspecies of the North American Savannah Sparrow, adapted to saltmarsh conditions, can process NaCl concentrations of up to 0.7 M, but these birds cannot be considered to be strictly arid-adapted. Species of birds not adapted to arid habitats cannot usually process fluids more concentrated than about 0.15 M NaCl.

5.7.2 Serum Osmolarity

The average osmolarity of the serum (= plasma) of most birds is around 390 mOsmol/l. Some desert birds can tolerate serum osmolarities of a little over 400 mOsmol/l, but most maintain osmolarities of about 300 mOsmol/l, even when dehydrated. Again the saltmarsh populations of the Savannah Sparrow are exceptional in allowing their serum to rise to osmolarities of over 600 mOsmol/l.

5.7.3 Urine Osmolarity

The concentrating abilities of avian kidneys and hindgut systems, while not as effective as those of mammals, allow the excretion of urine osmolarities of around 327 mOsmol/l on average, but this may rise in most species to between 800 and 1000 mOsmol/l when the birds are dehydrated or when they are drinking saline solutions. Desert-adapted species of birds are seldom better than this and can concentrate their urine to 500-600 mOsmol/l as a general rule. Species which inhabit arid regions, but which have free access to drinking water do not have special water-conservation measures and have urine osmolarities of about average. The saltmarsh populations of the Savannah Sparrow are again outstanding in this regard and can excrete urine concentrations of up to 2000 mOsmol/l.

5.7.4 Faecal Water Loss

Dehydrated birds can increase the resorption of water through the lower gut, including the caeca when present, to produce excreta (faeces and urine together) with a water content as low as 25-30%, but the average water contents of avian excreta, even in the most efficient species, are usually around 50-60%. Excreta of hydrated birds of all kinds have a water content of about 80%.

Birds that inhabit deserts are better at conserving water than those of better-watered habitats, but their osmoregulatory adaptations may be regarded rather as refinements of preadaptations than evolutionary end-products acquired under the selective pressure of a hot, waterless environment. Many of their water-conservation strategies are behavioural, rather than physiological, as will be shown in Chapter 6 on thermo-regulation.

6 Thermoregulation

Organisms which live in the deserts of the world may be subjected to some of the highest (and also the lowest) ambient temperatures (T_A) on the surface of the planet, but it is not merely the T_A itself which poses thermoregulatory problems: it is also the scarcity of shade and water. Animals in deserts cannot always therefore escape to a cooler microclimate in the heat of the day, and they must conserve their water resources as far as possible; this means that evaporative heat loss has to be balanced against the availability of water, as shown in the previous chapter. How birds meet these environmental and physiological challenges is the subject of the following account.

6.1 Body Temperature

Most birds have a body temperature (T_B) of 40-42 °C, which is somewhat higher than the average for mammals (1.87 °C higher at rest, 2.43 °C higher when active; Schmidt-Nielsen 1964). This may drop to 38.5 °C at certain times of the 24-h cycle. The day-night range of oscillation of T_B decreases with increasing body mass, and T_B decreases only slightly with decreasing body mass (Prinzinger et al. 1991). Body temperatures of birds living in hot environments do not differ significantly from those of birds in other environments, and their upper lethal temperatures appear to be similar (Dawson and Hudson 1970). An increase in T_B of about 4 °C appears to be tolerated by most birds, constituting moderate hyperthermia, which is regarded as "a cornerstone of avian adjustment to dry heat in particular".

The relatively high T_B of birds preadapts them to high T_A, since they can maintain a larger temperature gradient between T_B and T_A, can therefore more easily offload heat to the environment, and thus do not incur such great heat loads as mammals would do (Dawson and Bartholomew 1968). Nevertheless, desert birds do not have higher T_B than birds of more mesic habitats. The Domestic Fowl *Gallus gallus* can withstand T_B of 45 °C for several hours, but 46 °C is critical and 47 °C invariably fatal; this is a fairly typical performance among birds generally. The air in many desert regions may nevertheless reach T_A of 45-46 °C, at which the birds are able to survive, though they may undergo a temporary loss of coordination. Many birds can survive at 45 °C for long periods, whether they live in deserts or not: some arid-zone examples include Abert's Towhee *Pipilo aberti*, a resident of the Colorado desert of North America (a more detailed account of which appears below (Dawson 1954)), the American Kestrel *Falco sparverius*, Gambel's Quail *Callipepla gambelii*, the California Quail *C. californica* and the American Mourning Dove *Zenaidura macroura* (Fig. 4.5).

At T_A of 40 °C the quail, dove, towhee and kestrel showed T_B well above what was normal for the species. Since these birds are capable of losing about a third of their heat production by evaporative cooling, this rise in T_B is suggested to be possibly adaptive as a water-conservation measure, heat being lost by conduction and radiation instead of by evaporation (Schmidt-Nielsen 1964).

As already indicated above, birds can generally tolerate a heat load of up to 4 °C above normal T_B. This is confirmed by the fact that, in towhees at least, the relative importance of evaporation as a cooling mechanism decreases with increasing T_A (Table 6.1), the reverse of the situation in mammals (Dawson 1954). More recent work has shown that Rock Doves (Feral Pigeons) *Columba livia* can survive and breed successfully at T_A 60 °C for several hours a day, losing heat by cutaneous evaporation and not by panting (vigorous movements of the thoracic cage at 40-700 cycles/min, depending on the species, with accompanying buccal ventilation) or gular fluttering (pulsations of the hyoid apparatus in the floor of the buccal cavity, apparently at the resonating frequency of the system) (Arad et al. 1987). The birds maintained their T_B at 43 °C at T_A of 50 °C for long periods. Normally hydrated pigeons lost almost all their water by cutaneous evaporation, but dehydrated birds lost most of their water through the respiratory system because the birds resorted to panting as the major mechanism for heat loss. The advantages of cutaneous evaporative water loss (CEWL) over panting and gular flutter are that no extra metabolic heat is produced by the activity of the respiratory or buccal muscles, and that CO_2 washout in the lungs (alkalosis) is avoided.

Table 6.1. Percentage of total heat production lost by evaporation in two species of towhee. (After Dawson 1954)

	Ambient temperature (°C)					
Species	36	37	38	39	40	41
Abert's Towhee	44.6	41.3	39.3	36.1	35.9	35.2
Brown Towhee	30.5	30.9	30.2	29.3	27.6	26.9

6.2 Heat Loss

6.2.1 Evaporative Heat Loss (EHL)

Heat exchange between a bird and its surroundings is proportional to surface area, whereas heat storage is proportional to body mass and volume (Weathers 1981). Measurements of the thermoregulatory abilities of six species of heat-stressed birds, three of them from hot, dry habitats (Black-throated Sparrow *Amphispiza bilineata*, House Finch *Carpodacus mexicanus* and Gambel's Quail) and three from mesic habitats (Cassin's Finch *Carpodacus cassinii*, Red-winged Blackbird *Agelaius phoeniceus* and Common Quail *Coturnix coturnix*) showed that, above 30 °C EHL increased exponentially in all species, at any given temperature varying directly with body size, but that the EHL below 40 °C was greater in the mesic than in the arid-adapted species, suggesting that desert birds combine lower rates of evaporative water loss (EWL) with a greater sensitivity of EHL to changes in T_A. This system of control is ideally suited to deserts.

At the upper critical temperature (T_{UC}) the EHL varies with the magnitude of the gradient T_B-T_A (Weathers 1981). Larger birds begin to augment EHL substantially well within the thermoneutral zone (TNZ, i.e. before T_{UC} is reached), whereas smaller species defer such augmentation until T_{UC} is reached, relying on a larger T_B-T_A gradient to dissipate heat non-evaporatively. This strategy favours water economy in small birds and is therefore adaptive.

Evaporative cooling via the respiratory tract in birds is enhanced by increased movement of the air by panting or by gular flutter. In most birds studied panting begins at T_A between 42.5 and 43.5 °C, but at T_A above 40 °C most birds are unable to balance heat uptake with EHL and build up a heat load, perhaps as much as 2 °C above normal T_B. This elevates metabolic heat production, up to 75% of which is compensated for by evaporative cooling, at least at moderate to high RH. Evaporative cooling seems therefore to limit heat storage rather than to prevent it, except in the Ostrich *Struthio camelus* (Fig. 3.6) which can maintain T_B of 39.3 °C at T_A of 51 °C. The Ostrich evaporates moisture from the respiratory surfaces by increasing both tidal volume and breathing rate (Dawson and Bartholomew 1968). At low RHs, such as are found in deserts, evaporative cooling may dissipate all metabolic heat, even at T_A of 44-45.9 °C. At a more usual RH such T_A might cause physiological damage. Response to heat stress by changing abruptly from a low rate to high rate of breathing is found in the Ostrich, the Barn Owl *Tyto alba*, the Feral Pigeon and the Roadrunner *Geococcyx californianus*, a pattern analogous to that found in dogs.

Panting may be augmented by gular flutter, which has been observed in the following bird families: Pelecanidae, Sulidae, Phalacrocoracidae, Anhingidae, Ardeidae, Anatidae, Phasianidae, Cacatuidae, Strigidae, Pteroclidae, Columbidae, Cuculidae, Caprimulgidae and Coliidae. The frequency of the flutter appears not be related to T_A, probably because it is a resonating frequency which requires little energy to maintain, but the amplitude increases in proportion to increasing T_A. Gular flutter is energetically much less expensive than panting, requiring about 10% less heat dissipation than panting (Dawson and Bartholomew 1968). The well developed gular fluttering of nightjars (Caprimulgidae) and their capacity for evaporative cooling appear to be adaptations which fit them for incubating in very exposed sites for long periods of time (Dawson and Bartholomew 1968).

At high T_A the rate of respiratory movements during panting may rise to 200 or more movements/min at high T_A, whereas gular flutter may occur at rates in excess of 400-600/min (Schmidt-Nielsen 1964). However, the tidal volume of the air in the respiratory system decreases during panting: although respiratory frequency may increase tenfold, the tidal volume decreases in pigeons from 4.5 to 1.2 ml, whereas the minute volume is estimated to increase at most only three to four times. The rate of gular flutter may match the breathing rate or may be independent of it. The light and flexible gular region can operate at a lower metabolic cost than the heavier and more rigid thoracic cage (Schmidt-Nielsen 1964).

If the rate of gular flutter is independent of the rate of breathing, as in the Cattle Egret *Bubulcus ibis*, it can occur at 965 cycles/min, whereas breathing occurs at 44 cycles/min (Dawson and Hudson 1970). In nightjars (e.g. Poorwill *Phalaenoptilus nuttallii*, Common Nighthawk *Chordeiles minor*, Spotted Nightjar *Eurostopodus guttatus*), less than 20% of the heat is dissipated through evaporation at T_A 43 °C; in these birds, gular flutter appears to "overshadow other means of evaporative cooling", allowing the birds to lose as much as three times their heat production at high T_A (Dawson and Hudson 1970). The Poorwill also has an unusually low resting metabolic rate (RMR).

The evasion of heat by seeking shelter in rock crevices (e.g. Rock Wren *Salpinctes obsoletus*) and other shady places is common among birds, as well as the practice of soaring, by which birds escape to lower air temperatures and less severe secondary radiation.

Whether a bird pants or gular flutters probably makes little difference to the efficiency of evaporative cooling, but the increased heat output during panting must be balanced against its effectiveness in evaporative cooling. The T_A at which birds begin to pant is no different between desert and non-desert species – around 42-43 °C. Although vigorous, prolonged evaporative cooling may lead to alkalosis in most birds, the Ostrich is again exceptional: it can cool evaporatively from its respiratory system for as much as 8 h without becoming alkalotic (Dawson and Hudson 1970). The role of the air sacs and lungs in evaporative cooling appears to vary in birds of different kinds, but it remains to be studied in detail.

Evaporative water loss (EWL) is increased at high T_A by evaporating water from the mucosal surfaces and through the skin (Dawson and Hudson 1970). EWL increases far more markedly at T_A above thermoneutrality than does oxygen consumption (and the consequent production of metabolic water), thereby exacerbating the problem of maintaining water balance (Bartholomew 1972). Below the lower critical temperature (T_{LC}) oxygen consumption increases, but EWL remains almost unchanged; the smaller the bird the higher its T_{LC}, and the greater its mass-specific conductance, thus favouring its adaptation to high T_A. Indeed birds of tropical to subtropical deserts generally have high T_{LC}. For a diurnal bird of low-latitude arid zones, a small body size with high conductance and a high T_{LC} are therefore likely to be selected for as adaptations to high daytime T_A (Dawson and Hudson 1970). This may account for the large numbers of small birds in arid zones generally, though such adaptations would seem to be disadvantageous in winter when T_A in many deserts is often below 0 °C.

It is significant that, contrary to the case in small birds, most small desert mammals have a lower than average conductance, as well as a wide range of diel rhythms of body temperature and a low RMR (Lovegrove and Heldmaier 1994). These features are suggested to be adaptations to a wide range of diel fluctuations of T_A. Since most of these animals are nocturnal (therefore inactive by day) and fossorial, the risk of hyperthermia is low. Birds would seem to be at a relative disadvantage in being largely diurnal and not being able to escape into a cool burrow.

6.2.2 Non-Evaporative Heat Loss

Non-evaporative heat loss occurs by three processes (Whittow 1976):

1. Radiation in the form of electromagnetic waves from the body to parts of the environment at lower temperatures;
2. Conduction by the transfer of energy from one molecule to the next, without gross translocation of molecules;
3. Convection by the movement of molecules of air over and away from the skin.

At T_A lower than T_B, heat can be lost by conduction and convection from the lungs because the inspired air is warmed as it removes heat from the tissues over which it passes; much of this heat is lost from the body on expiration.

Various mechanisms are employed in order to lose heat to the environment by conduction and convection (Dawson and Bartholomew 1968):

1. Increasing the surface area by holding the wings away from the body;
2. Reducing insulation by compressing the feathers and exposing unfeathered parts of the body, such as the underwings and the legs;
3. Increasing blood flow to the legs or other bare areas.

Such adjustments may increase the thermal conductance of a small bird (about 40 g body mass) about three to four times and may be enhanced by further behavioural adjustments to augment physiological mechanisms which may be precarious at very high T_A. These behavioural adjustments include:

1. Seeking shade (reduced insolation and radiation);
2. Soaring to high altitudes (reduced radiation and lower temperatures);
3. Repeated bathing if water is available;
4. Minimizing activity.

The advantages of soaring may be further enhanced by flying with the legs dangling and acting as thermal windows.

6.3 Metabolic Rate

Variation in basal or resting metabolic rate (RMR) is strongly correlated with differences in body weight in birds when analyzed at the level of family (Bennett and Harvey 1987). (Resting metabolic rate is preferred to the more usual "basal metabolic rate" because not all experimental measurements of such rate are in fact minimal or "basal".) Larger animals have greater metabolic needs, but their metabolic rate does not increase in proportion to body size. Metabolic rate (*M*) increases with body weight (*W*) according to a power relationship, the exponent (*b*) of which is < 1 (its empirical limits are 0.66-1.00; Crawford and Lasiewski 1968):

$$M = aW^b.$$

After logarithmic transformation, *M* is linearly related to *W* and a is a constant (Fig. 6.1). The correlation between log RMR and log body weight is strong at ($r^2 = 0.99$).

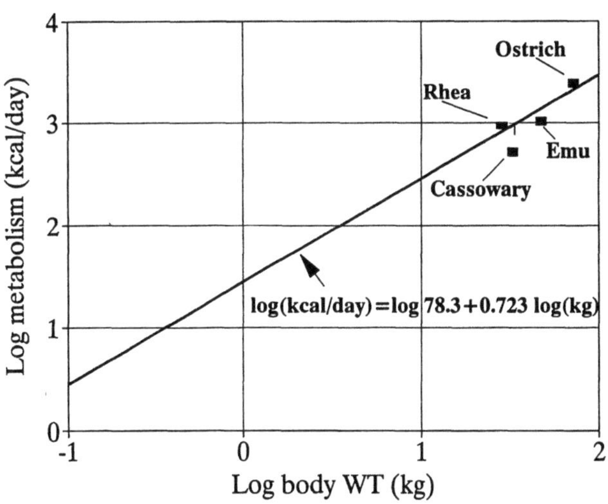

Fig. 6.1. Standard metabolic rate (SMR) relative to body weight in non-passerines (*straight line*) and some ratites. (Redrawn from Crawford and Lasiewski 1968). Kcal converted to kJ by multiplying by 4.184

The relationship between RMR and body weight is also strongly taxon-related. General conclusions are (Bennett and Harvey 1987):

1. Passerines have a higher RMR than non-passerines.
2. Significant differences exist between non-passerine taxa, so they should not be lumped together in an analysis.
3. Tropical birds have lower RMRs than temperate species.
4. Dark-coloured birds have lower relative metabolic rates than light-coloured birds.
5. Birds that forage in the open have lower relative metabolic rates than those that forage in shade.
6. Nocturnal birds have a relatively low RMR (possibly an adaptation to the higher T_A encountered by day when they are resting).
7. Lowering of RMR is adaptive in times of food shortage, or when building up food reserves for breeding.
8. Adaptive lowering of RMR in response to higher T_A might therefore increase chances of survival and reproduction.
9. It is important in interspecific studies of dietary correlates with RMR to attempt to establish whether associations arise from convergent evolution or from real phylogenetic relationship, so as to avoid revealing spurious relationships [this point has been extensively emphasized by Harvey and Pagel (1991), who highlight the interplay between phylogeny, chance and adaptation to contemporary environments].
10. Active metabolic rate (AMR) is significantly higher when measured in species undertaking breeding activities at the time.
11. Active, small-bodied birds require disproportionately more energy per unit time above resting level than do larger birds, probably because they spend more time flying.

The above overview will give some idea of the performances of birds under hot, dry conditions, and how they balance heat loss (especially by evaporation) with water balance. It is now worth looking at some individual case histories, from some of which the review material has been derived, in order to survey the adaptive performances of these birds in somewhat greater detail and then to draw a set of overall conclusions about the thermal adaptations of birds to arid zones (if any).

6.4 Some Case Histories

6.4.1 North American Towhees

Dawson (1954) made a thorough comparative study of two species of North American towhees (Fringillidae), one a desert-dweller, the other not. Abert's Towhee *Pipilo aberti* inhabits arid regions in North America where summer temperatures are high. The Brown Towhee *P. fuscus* inhabits more mesic regions, but penetrates to the edge of the desert, generally where summer temperatures are more moderate. At T_A 23 °C the mean T_B of the Brown Towhee was on average consistently lower than that of Abert's Towhee by day and consistently higher at night, but the differences were small and not statistically significant. Both were about 2 °C lower at night than by day, this lability being related probably to their small body size.

At T_A 39 °C, the T_B of both species was about 2 °C higher at night (i.e. about 41 °C) than at lower T_A, but by day only about 0.5 °C in the Brown Towhee (i.e. about 39.5 °C) and about equal (i.e. about 39 °C) in Abert's Towhee; indeed, the T_B in Abert's Towhee at T_A 39 °C was sometimes slightly lower than at T_A 23 °C. When deprived of water, the T_B of both species was up to 2 °C higher than at lower T_A, but the difference was generally less in Abert's Towhee than in the Brown Towhee. Abert's Towhee can tolerate T_B of 44-45 °C for short periods without ill effects. The mean lethal T_B for both species of towhee was 46.9 °C; above T_B 45 °C both species showed a temporary loss of coordination, even if they did not die. Generally the birds became inactive at high T_A.

Abert's Towhee has a wide TNZ of 25-35 °C. Above these values the RMR increases rapidly, and below these values it increases slowly but steadily. The TNZ for the Brown Towhee extends from 23-34 °C and is therefore slightly lower. Below this range the RMR increases much less steeply than in Abert's Towhee, but above the TNZ it increases in the same way as in Abert's Towhee, though more rapidly (cf. Figs. 6.2 and 6.3).

Under natural conditions of high T_A Abert's Towhee is active early morning and late afternoon. It becomes inactive during the hottest part of the day and retreats to the shelter of dense vegetation along watercourses, preferably where water is present. The Brown Towhee has a similar activity pattern in summer, but needs to move only short distances to find shade, because it inhabits denser vegetation.

The overall conclusions from this study are that Abert's Towhee tolerates heat somewhat better than does the Brown Towhee:

1. The coefficient of EWL of Abert's Towhee remains constant at high T_A, whereas that of the Brown Towhee increases.
2. During tests of EWL, the survival of Abert's Towhee is better than that of the Brown Towhee.
3. At T_A 39 °C for 24 h, the T_B of Abert's Towhee is slightly less elevated than that of the Brown Towhee.
4. At T_A 39 °C for 24 h without water, Abert's Towhee suffered no apparent ill effects, whereas several Brown Towhees died under these conditions.
5. At T_A 39.5 °C, Abert's Towhee drank less water per day than did the Brown Towhee.

Despite these conclusions, the capacities of Abert's Towhee for tolerating heat are limited and possibly merely a product of acclimation rather than a genetic ability. Abert's Towhee possesses no special adaptations for desert life, despite its somewhat better performance than the Brown Towhee's under experimental conditions. Survival of small passerines under hot conditions depends on a fine balance between heat storage and heat loss, as well as on

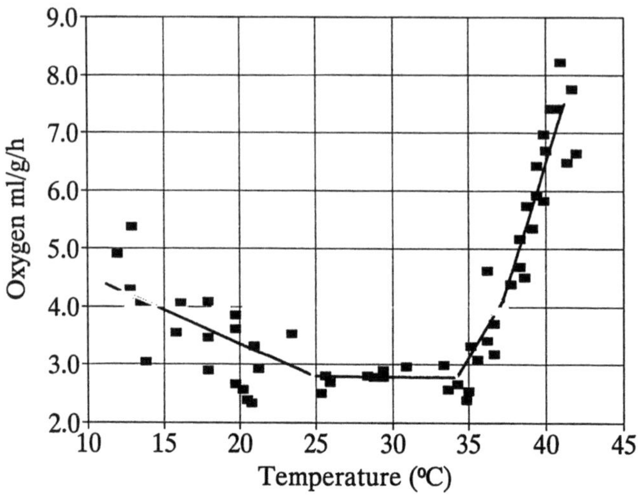

Fig. 6.2. Standard metabolic rate (SMR) in Abert's Towhee *Pipilo aberti* relative to ambient temperature. (Redrawn from Dawson 1954)

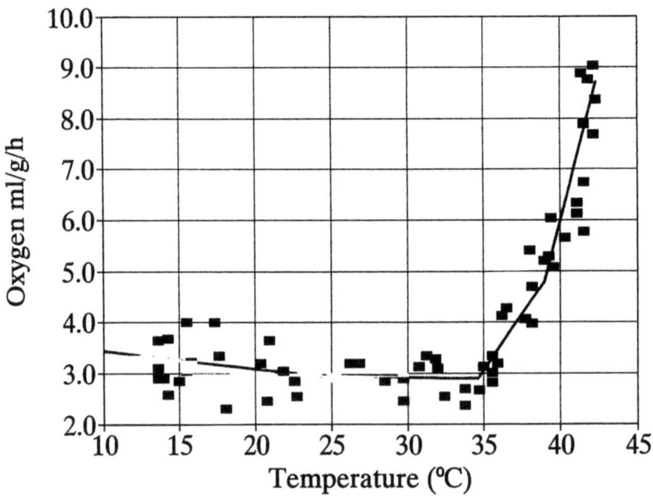

Fig. 6.3. Standard metabolic rate (SMR) in Abert's Towhee *Pipilo fuscus* relative to ambient temperature. (Redrawn from Dawson 1954)

behaviour affecting the production of heat and the gain of heat from the environment. Since in deserts severe heat stress lasts for only a few hours each day, "the basic problem of desert birds exposed to high air temperatures may be regarded as one of retarding the rise of body temperature until environmental conditions ameliorate, more than one of maintaining a steady state of body temperature" (Dawson 1954). The task is made more difficult for most passerines because of their small body size.

Table 6.2. Mean body weights, conductance and temperature coefficients of some selected passerines of different habitats. (Calder 1964)

Species	Mean body weight (g)	Conductance (kJ/°C/m²/h)	Temperature coefficient
Red-capped Manakin *Pipra mentalis*	12.0	10.71	NA
Grey Jay *Perisoreus canadensis*	64.0	4.98	NA
Blue Jay *Cyanocitta cristata*	80.8	7.32	NA
House Wren *Troglodytes aedon*	10.8	NA	4.2
House Sparrow *Passer domesticus*	25.5	NA	2.0
Zebra Finch *Poephila guttata*[a]	11.7	14.43	7.9
Paradise Whydah *Vidua paradisaea*	10.5	13.81	5.6
Abert's Towhee *Pipilo aberti*[a]	46.8	11.63	3.8
Brown Towhee *Pipilo fuscus*	43.7	10.71	1.8
Slate-coloured Junco *Junco hyemalis*	19.0	NA	2.0
Snow Bunting *Plectrophenax nivalis*	58.0[b]	8.70	NA

NA, not available
[a]Desert inhabitants.
[b]Almost certainly an error for "28.0", considering the size of the bird.

6.4.2 The Zebra Finch *Poephila guttata*

In a study of the Zebra Finch Calder (1964) showed that motor activity is minimized when the birds are heat-stressed and that they tolerate limited hyperthermia rather than use evaporative methods of heat loss: T_B rose from 41 °C in the TNZ to a maximum of 43-43.5 °C above TNZ, at which point evaporative cooling was resorted to, but without the advantages of reduced RMR and gular flutter of extensive oral surfaces as in the Poorwill (Bartholomew et al. 1962).

The Zebra Finch has thin plumage and a high conductance for heat loss from the body. Conductance is measured by dividing the metabolic rate by the temperature gradient at T_{LC} and is expressed as kJ/°C per m² per h. Another measure of insulative properties is the temperature coefficient, an expression of the metabolic increase required to maintain T_B as T_A decreases, expressed as the percentage change of RMR at T_{LC} per °C drop in T_A. A comparison of these values for a few species of birds is shown in Table 6.2.

As T_A overtakes T_B, loss of metabolic heat is no longer possible by passive means. At T_A of about 43 °C, prevention of harmful hyperthermia becomes more important than conserving water, so that evaporative cooling is used. When given water *ad libitum* the T_B was a mean of 43.1 °C at a mean T_A of 44.2 °C, but when deprived of water at a mean T_A of 43.9 °C the mean T_B was 43.4 °C. T_B was therefore maintained at about 3 °C below the upper lethal temperature, but the birds became exhausted and dehydrated, indicating the evaporative cooling under such conditions is an emergency measure.

6.4.3 Passerines in the Namib Desert

Most bird species in the Namib remain relatively inactive during the hot hours of the day, usually seeking shade (Willoughby and Cade 1967). The Dune Lark *Mirafra erythrochlamys*, for example forages until the sand reaches about 50 °C, then shelters in the tufts of *Stipagrostis sabulicola* until the sand temperature again drops below 50 °C when it emerges to forage in the late afternoon. Gray's Lark *Ammomanes grayi*, the Tractrac Chat *Cercomela tractrac* and some other small ground birds avoid contact with the hot ground by perching a few centimetres above ground level on stones or twigs, which also exposes them to the relatively cool (28-32 °C) sea breeze which often blows across the Namib flats in the afternoon. The birds face into the airstream while holding the wrists out from the sides of the body to expose the poorly feathered regions under the wings and on the flanks in order to lose heat by convection.

6.4.4 The Cactus Wren *Campylorhynchus brunneicapillus*

The Cactus Wren is a permanent resident of the arid Lower Sonoran Desert. It is insectivorous and lives in regions without free water; it does not drink even when free water is available (Ricklefs and Hainsworth 1968). Foraging activity varies with T_A. Foraging activity by adults and their feeding of young is depressed at T_A>35 °C when only the coolest microhabitats are used. The diversity of foraging location decreases with increasing T_A. Greatest foraging activity occurs at T_A<32 °C.

Fig. 6.4. Nest mass of Sociable Weaver *Philetairus socius*, Kalahari Desert, South Africa

6.4.5 The Sociable Weaver *Philetairus socius*

The interior of the chambers of the nest of the Sociable Weaver (Fig. 6.4) remains well within the zone of thermoneutrality for a passerine bird of this size (about 27 g; Maclean 1993): it is always above T_A at night and below T_A by day (Bartholomew et al. 1976). In winter up to five birds roost in a single chamber, but in summer no more than two birds roost in a chamber. "The principal mechanism for maintaining chambers within the zone of minimal energetic cost is changes in the number of birds in the nest chambers at night" (Bartholomew et al. 1976). Humidity in the chambers is always less than that outside because the nest material takes up most of the water vapour from the air filtering through the nest mass; this moisture is dissipated during the day, so that the nest materials remain dry.

6.4.6 Estrildids

Marschall and Prinzinger (1991) subjected five species of estrildids to the following calculation from laboratory measurements under different T_A: O_2 consumption, CO_2 production, RMR, TNZ, conductance and T_B. The species were (1) Cut-throat Finch *Amadina fasciata*; (2) Orange-cheeked Waxbill *Estrilda melpoda*; (3) Zebra Finch *Poephila guttata*; (4) Gouldian Finch *Chloebia gouldiae*; and (5) Java Sparrow *Padda oryzivora*. Results are shown in Table 6.3.

The RMR (given as BMR in Table 6.3) in all species accords with the expected RMR for birds of their weights in tropical habitats and is lower than that of birds of equivalent mass in mesic climates. The advantages are: (1) less energy used and therefore less food needed; (2) the time for foraging is reduced and therefore the birds do not need to be in the sun for so long; and (3) lowered O_2 consumption means reduced rates of respiratory water loss.

The TNZ of all five species has a relatively high T_{UC}, and in some cases the T_{UC} was not even reached at 42 °C. This is an adaptation to high T_A, whereby heat production and evaporation are minimized. This is especially important for desert species. Arid environments also undergo huge daily fluctuations in T_A from day to night, so that the building of roosting nests by the arid-zone Cut-throat Finch and the Zebra Finch (and by several other species of birds also, notably many other estrildids and the Scaly-feathered Finch) is a mechanism to cope with low night-time temperatures (and the birds' high thermal conductances).

If conductance is low, insulation is likely to be high and the bird is likely to be adapted to low T_A. However, under conditions of high T_A, conductance should be high to facilitate heat loss, as is the case in the five estrildids under study. This economizes on evaporative water loss for thermoregulatory purposes. There is also a relationship between RMR and conductance ("Con" in Table 6.3), in that both are relatively low (as in the Cut-throat Finch) or both are relatively high (as in the Zebra Finch).

Table 6.3. Mean physiological parameters of five species of estrildid. (Marschall and Prinzinger 1991).

Parameter	Species				
	Cut-throat Finch	Orange-cheeked Waxbill	Zebra Finch	Gouldian Finch	Java Sparrow
T_B	43.1	43.0	43.5	42.1	42.5
BMR	44.8	62.5	54.7	50.5	43.8
%Δ	-26.5	-18.4	-17.9	-19.6	-20.1
Con	4.0	6.2	5.2	4.2	3.4
%Δ	+25.0	+34.8	+44.4	+31.3	+30.8
TNZ	31-38	35-42[a]	33-38[a]	30-42[a]	31- 42[a]

T_B, Daytime body temperature; BMR, basal metabolic rate (J/g per h); %Δ, % difference from that of typical bird; Con, conductance (J/g per h per °C); TNZ, thermoneutral zone (°C).
[a] Upper critical temperature not reached.

Fig. 6.5. Emu *Dromaius novaehollandiae*, New South Wales, Australia

Body temperature of all five species was relatively high for passerines (42.1-43.5 °C), and rose only slowly with increasing T_A. At no time did any of them reach proper hyperthermia. Because of the lowered metabolism and high conductance, an increase in T_B for the purpose of heat loss is unnecessary. Even though the Cut-throat Finch attained its T_{UC}, it did not display hyperthermia, and underwent only a slight rise in T_B.

6.4.7 Ratites

The T_B of all the larger ratites is lower than that for most small birds. The standard metabolism for the Emu *Dromaius novaehollandiae* (Fig. 6.5; weight up to 55 kg; mean T_B 38.1 °C) and the Greater Rhea (or Ñandú) *Rhea americana* (weight about 20 kg; mean T_B 39.7 °C) is similar to that expected for large non-flying birds of this size (Crawford and Lasiewski 1968). A rate of EWL of 179 mg H_2O/min was measured for the Emu and 160 mg/min for the Rhea, representing 0.67 and 1.07% of body weight per day respectively. This is primarily respiratory water loss, but some comes from cutaneous evaporation in the head and neck regions. The mean resting respiratory and heart-beat rates for the Emu (7 breaths/min; 41 beats/min), the Rhea (9 breaths/min; 48 beats/min) and the resting respiratory rate of the Ostrich *Struthio camelus* (Fig. 3.6; 4-7 breaths/min) are lower than corresponding rates for mammals of comparable size. Since the relationship between body size and metabolic rates of mammals and non-passerine birds does not differ significantly, the differences between the respiratory and heartbeat rates of mammals and birds raise some puzzling questions of avian circulatory physiology (Crawford and Lasiewski 1968).

Louw et al. (1969) made a study of thermoregulatory physiology of the Ostrich, in which they measured T_B and respiration rate. The range of T_B in the Ostrich is generally between 38.3 °C and 39.8 °C, measured by probes implanted into the left gluteal muscle to a depth of 5 cm. The lowest and highest T_B measured in this way were respectively 37.9 °C and 40.7 °C. T_B varies diurnally on a cyclic basis, being maximal between 12:00 and 17:00 h and minimal around 05:00 h. The temperature in the air space between the skin and feather layer can be controlled and must play an important role in thermoregulation: when T_B approaches the upper limits of the diurnal cycle (around 40 °C) the air-space temperature drops. This decrease is aided by raising the plumage at high T_A, allowing convective cooling to occur.

The rate of respiration fell into two class intervals: 3-5 respirations/min and 36-47 respirations/min with no intermediates. As T_A increased to a critical threshold of 33-36 °C, the rate of respiration increased suddenly, within seconds, from the lower to the higher rate. In the presence of a breeze, this critical temperature may not be reached, illustrating further the importance to the Ostrich of convective cooling. The rapid rate of respiration serves the purpose of evaporative cooling. Even at high respiration rates the Ostrich does not undergo alkalosis, the suggestion being that under high T_A a functional shunt may exist in which air bypasses the lungs and circulates increasingly through the air sacs (Schmidt-Nielsen et al. 1967).

In addition to the erection of plumage at high T_A, Ostriches also droop the wings forward and downward from the body, thereby exposing the unfeathered thorax and thighs (Fig. 6.6), facilitating further heat loss by convection and radiation through these thermal windows. At night when T_A dropped considerably, the wings and plumage were held close to the body to provide insulation (Schmidt-Nielsen et al. 1967).

Fig. 6.6. Ostrich *Struthio camelus* drooping wings and exposing naked tibiotarsus for heat loss, South Africa

The Ostrich can maintain its T_B below 40 °C for at least 8 hours at T_A as high as 50 °C, losing water at only about half the rate as a man sweating under similar conditions (Schmidt-Nielsen et al. 1969). Respiratory rates rose from 6-12 cycles/min at rest to 40-60 cycles/min under heat load. Water loss through the skin "was apparently insignificant" (but see Bernstein 1971 in Chap. 5: he found "significant" CEWL in several species of smaller birds, both passerine and non-passerine).

The Ostrich can reduce the uptake of heat by erecting the plumage in order to increase the thickness of the insulating layer (Wiens 1991). Sandgrouse achieve the same result with their thick undercoat of brown down. Most thermoregulation is achieved, however, by evaporative cooling. This evaporation may take place through the skin, by panting or by gular flutter. (These mechanisms are not unique to desert birds.) In the dry air of deserts these mechanisms allow the complete dissipation of excess heat in most birds at T_A up to 45 °C. The Ostrich and the Spotted Nightjar of Australia can maintain their normal T_B at T_A of 50-52 °C (Dawson and Fisher 1969; Schmidt-Nielsen et al. 1969).

6.4.8 Sandgrouse (Figs 3.8 and 3.9)

Sandgrouse are typically relatively small ground-dwelling birds which are exposed to high T_A and low RH (Marder et al. 1986). The desert and steppe habitats of Morocco impose extreme thermal conditions on sandgrouse which are diurnally active and nest in the open. T_A may reach 55 °C (Thomas and Robin 1977). Metabolic and environmental heat loads are kept to a minimum by concentrating vigorous activity to the cooler parts of the day, becoming inactive in the heat and by increasing thermal insulation when T_A exceeds T_B. Evaporative cooling by gular flutter and gaping is delayed until T_A reaches about 50 °C. Sandgrouse are unwilling to fly in the heat of the day; if flushed they fly only short distances before alighting. The Chestnut-bellied Sandgrouse *Pterocles exustus* in India and Pakistan rests in the shade of small shrubs during the heat of the day (Christensen et al. 1964).

The behaviour of a female Pin-tailed Sandgrouse *Pterocles alchata* on the nest showed that feather erection occurred as T_A approached 30 °C and increased to a maximum when T_A reached 50 °C (Thomas and Robin 1977). Crowned Sandgrouse *P. coronatus* and Spotted Sandgrouse *P. senegallus* did not "ruffle" their dorsal plumage at high T_A as obviously as *P. alchata*, but they looked plump and could have been increasing their insulation without separating the tips of the dorsal feathers, as happens in "ruffling". At nearly 50 °C gular flutter began and when it exceeded 50 °C gaping accompanied gular flutter until T_A dropped to 45 °C. Sandgrouse did not appear to pant. No sandgrouse on the nest appeared to align the body with respect to the sun in order to minimize the solar heat load.

Increasing feather insulation at high T_A is not unique to sandgrouse as claimed by Thomas and Robin (1977): it has also been observed in the Ostrich (Louw et al. 1969), and almost certainly occurs in other birds too. Sandgrouse lack areas of bare skin through which to lose heat, possibly because they are often exposed to T_A well above normal T_B, when evaporative cooling takes place. The feather insulation appears to be very effective in retarding an increased T_B at high T_A, but "since insulation necessarily retains metabolic heat, it would appear that sandgrouse also tolerate internal hyperthermia" (Thomas and Robin 1977). (I would suggest that it is possible that sandgrouse can lower their metabolic rate in order to keep metabolic heat production to a minimum.)

Cloacal temperatures of freshly killed Namaqua Sandgrouse *Pterocles namaqua* varied from 41.5-42.0 °C (Dixon and Louw 1978). Newly hatched chicks were thermolabile and not fully able to thermoregulate. Incubating adults on the nest oriented themselves facing into the sun in order to minimize the exposed surface area, but if there was a breeze they faced into it, especially if it was a cool sea breeze, holding the wrists away from the body to aid convective cooling; the whole body may be raised from the eggs at such times, in order to allow the eggs to cool also, while remaining shaded by the body of the parent. At high T_A the parent bird might also raise the feathers of the upper back in order to optimize convective cooling. Gular flutter was employed for evaporative cooling.

Namaqua Sandgrouse and Double-banded Sandgrouse *P. bicinctus* were studied in the field and in captivity (Thomas and Maclean 1981). Under T_A >30 °C they drooped their wings, sought shade and reduced activity; at T_A >40 °C they huddled together and began to gular flutter. They withstood water deprivation for 3 days without ill effect, but intensified some of their thermoregulatory mechanisms. Cloacal temperatures varied from 39 °C at night to 41 °C by day and the birds showed no hyperthermia at T_A >50 °C. When dehydrated, their T_B was reduced by 1.1 °C. Heat storage by the birds resulted in a saving of water by 2-5% when they had free access to drinking water, but by as much as 7-10% when water-deprived. The metabolic heat production, calculated from the food eaten, was 103-119% of the resting metabolic heat production for other kinds of birds of similar body mass, suggesting a low RMR in sandgrouse, resulting in low water turnover. Sandgrouse therefore appear to be adapted to arid environments by conserving energy and water, and by being able to withstand high T_A by means of several behavioural and physiological mechanisms.

Namaqua Sandgrouse and Double-banded Sandgrouse seek shade in the heat of the day and tend to face the sun (Thomas et al. 1981). Huddling in captive birds occurs at night and in the heat of the day. (My own field observations in the Kalahari, however, failed to detect huddling in wild Namaqua Sandgrouse in the heat of the day.) Gular flutter is not used at all by dehydrated captive birds. Sandgrouse have a relatively high standard metabolic rate (SMR), contrary to the conclusions of Thomas and Maclean (1981), and need an effective mechanism for dissipating heat (Marder et al. 1986). The Spotted Sandgrouse is a successful inhabitant of the desert and incubating females do not gape or gular flutter under heat stress. This indicates that cooling of the body and the eggs could be achieved only by efficient cutaneous evaporation of water (CEWL). This is a mechanism that has been demonstrated in other species of birds, and is important for survival under conditions of extreme heat.

Sandgrouse exposed experimentally to T_A from 27-51 °C for 2-3 h had T_B varying from 41.4-42.6 °C (Thomas et al. 1981). At T_A 51 °C and RH <10% for 240 min the skin temperature (T_S) remained stable at 42 °C. The T_S was always lower than T_B by about 0.5-1.1 °C at any T_A, but the difference between T_B and T_S decreased as T_A increased. The same has been found in the Feral Pigeon *Columba livia*. High-frequency panting and gular flutter were recorded at T_A of 40-60 °C, during which time the experimental birds remain very still, but show no signs of distress.

Table 6.4. Mean body temperature (T_B, °C), skin temperature (T_S, °C), cutaneous resistance to vapour diffusion (R_S, s/cm), and cutaneous water loss (CEWL, mg H_2O/cm^2 per h) in the Spotted Sandgrouse *Pterocles senegallus* experimentally exposed to high ambient temperatures (T_A, °C) after 5 weeks of acclimatization. (Marder et al. 1986)

T_A	T_B	T_S	$T_B - T_S$	R_S	CEWL
27	41.4	40.3	1.1	103.0	1.6
42	42.0	41.3	0.7	20.5	11.1
45	42.4	41.6	0.8	13.5	14.3
51	42.6	42.1	0.5	15.0	13.6

The rate of CEWL was 1.6 mg H_2O/cm^2 per h at T_A 27 °C with a cutaneous resistance to vapour diffusion (R_S) of 103 s/cm. When heat-stressed at T_A of 41-51 °C the R_S decreased to 20.5-15.0 s/cm and the CEWL increased to 13.6 mg H_2O/cm^2 per h at the maximum T_A (Table 6.4). Since CEWL was measured on unfeathered pectoral skin, the actual value may be less when measured on feathered skin where the water has to traverse the feathers and the boundary of air trapped between them and the skin.

While hyperthermia may indeed be a mechanism for withstanding high T_A, it is suggested that it may be merely a consequence of an inability to cool sufficiently by dissipating external and metabolic heat by evaporation. The absence of marked hyperthermia in the Spotted Sandgrouse supports the view that CEWL in sandgrouse (and pigeons) of the desert is an important mechanism for surviving hot environmental conditions (Marder et al. 1986).

6.4.9 Columbids

Bartholomew and Dawson (1954) showed that the American Mourning Dove *Zenaidura macroura* (Fig. 4.5; body weight about 150 g) had a labile T_B. The mean T_B by day was 41.5 °C, but it varied between 39.4 °C and 43.3 °C, depending on variation in activity. Under heat stress the mean T_B was higher than that characteristic of the TNZ, especially at night. If deprived of water the dehydrated birds had an even higher T_B (up to about 44 °C), but the birds maintained this condition even under prolonged high T_A and no water.

The thermal resistance of plumage in experimental black and white Feral Pigeons was 39% (with feathers depressed) and 16% (with feathers erected) of the thermal resistance of still air of equivalent depth (Walsberg et al. 1978). Erection of the feathers increased the depth of the plumage fourfold and the thermal resistance of the plumage by about 56%. At low wind velocities black plumage acquired a greater heat load than white plumage, but radiation penetrated white plumage more easily so that its radiative heating was less affected by convective cooling than that of black plumage. Therefore, as wind velocity increased, the heat loads of black and white plumages converged, especially when the plumage was erected. Thus, animals with dark coats may acquire smaller heat loads than pale animals under naturally hot conditions. Perhaps it is for this reason that the commonest non-cryptic colour among desert birds is black (conversely, it is probably why many polar animals are white).

Using artificially bred white Zebra Finches, some of which were experimentally dyed black, Hamilton and Heppner (1967) showed that the birds with black plumage had an average metabolic economy of 22.9% over the birds with white plumage, and that dark pigmentation facilitates the process of absorption and utilization of radiant solar energy. This may be important for animals living in climates in which the daytime temperatures may be high, but the temperatures at dawn and dusk may be low: the ability to absorb solar radiation at such times is an important economy measure, allowing the organisms to save on metabolic energy to maintain T_B (Hamilton and Heppner 1967). These workers indicated further that such a capacity is likely to be more important in the metabolic economy of an animal that the "slightly greater emissivity" conferred by dark coloration. Since wide ranges of T_A are characteristic of deserts, dark plumage in birds would be an advantage, but this must be weighed against the more important need of most ground birds to be cryptically coloured in order to reduce the chance of predation. Once again therefore, it seems as if there is good reason for black to be the most prevalent non-cryptic colour of desert birds.

Selective heating of the spinal cord of pigeons evokes panting. Cooling the spinal cord stops panting elicited by ambient temperatures of 38 °C (Rautenberg et al. 1978). Spinal receptors appear therefore to be the logical site for sensing ambient temperature changes; stimulation of these receptors governs temperature-related behaviour.

The Plumed (or Spinifex) Pigeon *Geophaps plumifera* inhabits semi-arid and arid regions of Australia with rocky and stony substrates. It flies relatively weakly and cannot travel long distances to water; it is therefore found near permanent water and drinks intermittently. Its food is the seeds of grasses, especially spinifex *Triodia* (Withers and Williams 1990). It experiences considerable water and heat stress in its environment, which includes some of the hottest places on earth. It weighs about 90 g, has a daytime T_B of 40.5-41.8 °C and a TNZ of about 35-45 °C. Its basal metabolic rate (= resting metabolic rate, RMR) is 0.85 ml O_2/g per h, which is about 68% of the predicted value for a non-passerine and 75% of the rate expected for a pigeon. The RMR increases below and above the TNZ. EWL increases exponentially with increasing T_A from < 1 mg/g per h at 0 °C to > 20 mg/g per h above 45 °C (Withers and Williams 1990).

Withers and Williams (1990) have shown that the Plumed Pigeon has remarkable thermal tolerance to high T_A because of its physiological characteristics:

1. Its metabolic heat production is low.
2. It becomes hyperthermic (T_B = 43.4 °C) at T_A >40 °C, which facilitates EHL.
3. Its dry thermal conductance increases 3-fold to 5-fold at T_A >30 °C, which facilitates non-evaporative heat loss.
4. EHL dissipates more than 100% of its metabolic heat production at T_A >40 °C.
5. Because the temperature of expired air is much lower than T_B at low T_As, respiratory evaporative water loss (REWL) is reduced to only about 20% of the total EWL, the remaining 80% of which is cutaneous (CEWL), but whose control is not clearly understood.
6. CEWL dissipates as much as 40% of metabolic heat production at high T_A.
7. Even at T_A of 45 °C, it does not appear distressed and does not gular flutter more than occasionally.

The Australian Diamond Dove *Geopelia cuneata* (mean body mass = 35 g) has a TNZ of about 34-46 °C, over which its metabolic rate remains at the basal level (Schleucher et al. 1991). Such a wide TNZ is an excellent adaptation to hot, dry conditions. A TNZ with such a high T_{UC} has been found in only a few other species of birds, all of them inhabitants of hot and/or dry environments: for example the Plumed Pigeon has a T_{UC} of 44 °C (Dawson and Bennett 1973; Withers and Williams 1990) and sandgrouse have a T_{UC} of up to 46 °C (Thomas 1984a, b). The Diamond Dove's metabolic rate increases during the daytime to 128% of the nighttime rate. The RMR of the Diamond Dove is 11% higher than predicted for a non-passerine of its body size (Bennett and Harvey 1987), but it is often active in the hottest time of the day, foraging in open sunshine. The suggestion is that it may be related to competition and predation in a habitat of low productivity, such as a desert.

At T_As from 14-36 °C, the mean daytime T_B of the Diamond Dove is stable at around 39.8 °C. At T_A <14 °C, T_B decreases only slightly to a minimum of 37 °C, but when T_A rises to >36 °C, T_B rises significantly to a maximum of 44.8 °C without any obvious harm to the bird. This ΔT_B of 5 °C represents a maximum heat storage of about 50% of 1 hour of metabolic heat production in the TNZ, and confers a water saving of up to 236 mg per bird. The Diamond Dove shows an impressive tolerance of high T_B at elevated T_A, an energy-saving mechanism which allows an increased difference between T_B and T_A, which in turn promotes passive non-evaporative heat loss, thereby further saving water. Keeping the T_B at lower temperatures would require extra energy which would produce more heat and necessitate further evaporative loss of water, an obvious ecological disadvantage.

Within the TNZ, EWL remains fairly constant at about 3.00 mg of water/g/h, but it rises exponentially to 20.56 mg/g per h at the maximum measured T_A. At T_A of 44 °C the amount of heat dissipated by evaporation is equal to that produced by the bird's metabolism. At T_A >32 °C regular breathing alternates with panting, but at T_A >40 °C gular flutter takes over.

The Diamond Dove has a thermal conductance higher than predicted (20% higher at night and 6% higher by day), in accordance with the suggestion that desert birds have high conductance which facilitates heat loss by non-evaporative mechanisms.

Immediately after drinking between 11:00 and 14:00 h Diamond Doves return to the bush along the dry riverbed to begin feeding again (Schleucher 1993). Feeding continues even when the T_A is maximal between 14:00 and 15:00 h, but intensifies towards evening. Other ground-foraging bird species do not feed at this time of day. Diamond Doves and Plumed Pigeons are often exposed to direct sunlight in the heat of the day and experience T_A above 40 °C for as much as 10 h of the day. They may rest on the ground for an hour at a time, where the temperature is as high as 62-65 °C, without showing any signs of distress. Diamond Doves are never entirely inactive during the daylight hours and engage in about 26% of their activity between 12:00 and 14:00 h without showing even the slightest signs of heat stress. Zebra Finches on the other hand spend only about 6% of their activity during those hours, and raptors only about 4%.

6.4.10 Charadriiformes Other Than Sandgrouse

At high T_A the Doublebanded Courser *Smutsornis africanus* (Fig. 2.3) pants and raises the dorsal plumage, especially if incubating in full sun (Maclean 1967b). Non-breeding birds seek shade. Even on the hottest days, adults show no signs of distress in the heat, but chicks rapidly collapse if exposed to the sun for even a short period.

The Inland (Australian) Dotterel *Peltohyas australis* (Fig. 3.1) is active mainly at night (Maclean 1976b). During the day it is largely inactive, but does not seek shade until T_A exceeds 40 °C. At T_A 46 °C all the birds seek shade (standing or crouching, depending on the height of the vegetation), pant and are reluctant to move into the sun, even when forced to do so by disturbance. The birds do not appear to use gular flutter.

Adult Australian Pratincoles *Stiltia isabella* (Fig. 4.7) do not seek full shade even when T_A reaches 46 °C, possibly because visibility may be impaired by so doing, but they may shade the legs by standing next to a shrub (Maclean 1976a; this could be an incidental consequence if the bird is merely seeking a cooler place to stand when the desert floor gets very hot). On hot days the birds walk over the gibber desert (claypan with a pavement of rounded stones; Fig. 1.6) with a slow, high-stepping gait holding the raised foot momentarily in the shade of the body before putting it down for the next step. Panting may start at T_A as low as 22.8 °C. When T_A reaches 25 °C in sunny weather the dorsal plumage is raised, either to increase the dissipation of heat by convection, or to increase the insulation against insolation; the intensity of panting also increases, but gular flutter is not employed. Birds may stand on a raised object where the temperature may be several degrees lower than on the desert floor, but this behaviour also occurred in mild weather, so it may simply be for enhanced vigilance.

6.4.11 Galliformes

Chukar Partridges *Alectoris chukar* cannot tolerate T_A of more than 42.5 °C, growing restless under experimental conditions, whereas Sand Partridges *Ammoperdix heyi* can tolerate T_A of 45 °C and maintain T_B below this level, whether watered or dehydrated (Kleinhaus et al. 1985). The T_Bs of both species begin to increase at T_A of about 35 °C, but the T_Bs of the two species show no significant differences from each other under different temperature conditions, whether watered or not. The same is true of brain temperature (T_{br}) of the two species. Water-deprived birds of both species breathe at significantly lower rates than watered birds; dehydrated birds pant and gular flutter at significantly higher T_A than watered birds of both species, but Sand Partridges begin to pant at a higher T_A (42.5 °C) than do Chukars (38.5 °C). The conclusion is that the Sand Partridge is better adapted to a hot, arid environment than is the Chukar.

Both species were shown to have lower standard metabolic rates (SMR) than would be predicted on their body masses. Their T_{LC}s were about the same at 28 °C for Chukars and 30 °C for Sand Partridges, but their T_{UC}s differed markedly: 38.5 °C for Chukars and in excess of 51 °C for Sand Partridges (Figs. 6.7 and 6.8). The Sand Partridges clearly thermoregulate better than Chukars at high T_A, which adapts them better to their more extremely arid habitats (Frumkin et al. 1986) and confirms the conclusion already stated earlier.

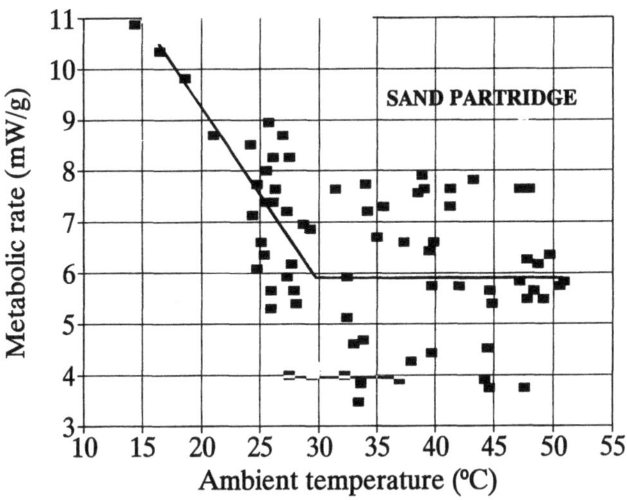

Fig. 6.7. Specific metabolic rate (expressed as mW/g) in the Sand Partridge *Ammoperdix heyi* relative to ambient temperature. (Redrawn from Frumkin et al. 1986)

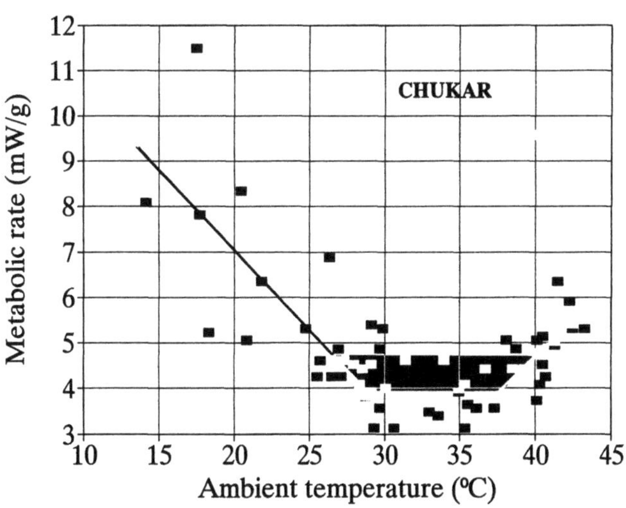

Fig. 6.8. Specific metabolic rate (expressed as mW/g) in the Chukar Partridge *Alectoris chukar* relative to ambient temperature. (Redrawn from Frumkin et al. 1986)

6.4.12 Cuculiformes

The Roadrunner (Fig. 5.6) sometimes suns itself in the early morning and intermittently on cold clear days, but never when TA exceeds its TNZ (Ohmart and Lasiewski 1971). Sunning is performed with the back orientated at right angles to the incident sunlight, while the black skin of the interscapular apterium and the soft black down of the dorsal spinal feather tract are exposed by raising the dorsal contour feathers. Roadrunners possess a number of adaptations to a desert existence, namely: (1) a salt-secreting nasal gland; (2) an ability to conserve water when dehydrated; and (3) an ability to conserve energy at low T_A through hypothermia and solar-heat absorption.

Roadrunners reduce their activity during the hottest hours of the day by slightly less than half. During the midday hours the birds divide their time between hunting for food and standing in the shade to cool off, at the same time gaping, gular fluttering and holding the wings away from the body (Calder 1968). This pattern of activity is similar to that of many birds in hot climates, and suggests that birds rely more upon behavioural mechanisms than upon physiological mechanisms for surviving desert conditions.

6.4.13 Caprimulgiformes

The Spotted Nightjar of Australia weighs 85-90 g. Its RMR is about 0.83 ml O_2/g per h, which is about 63% of the predicted value for an 88-g non-passerine bird (Dawson and Fisher 1969). This low metabolic level agrees with those found in two other caprimulgiforms, the Poorwill (40 g, 49% of predicted) (Bartholomew et al. 1962) and the Common Nighthawk *Chordeiles minor* (75 g, 79% of predicted) (Lasiewski and Dawson 1964). Oxygen consumption increases relatively little at high T_A, even above 50 °C, but prolonged exposure to 56.5 °C is fatal, probably because efforts to cool evaporatively increase the heat load. Gular fluttering reduces the energy requirements of evaporative cooling; in the Poorwill gular flutter starts only at about 39 °C, its frequency being independent of heat load (Lasiewski and Bartholomew 1966). Adjustments to the heat load are made by varying (1) the time spent gular fluttering; (2) the amplitude of movements of the gular region; and (3) the gular area exposed. The gular region of the Poorwill may be maintained at 3 °C below T_B and as much as 9 °C below T_A. The evaporative cooling mechanisms of the Spotted Nightjar dissipate rather more than three times the bird's metabolic heat production at high T_A, which is extremely efficient for a bird inhabiting hot climates. Spotted Nightjars obtain water from their insect food and they drink on the wing.

6.5 Conclusions

In general, desert birds do not appear to have any special structural or physiological adaptations to the arid environment, except that their tolerance to prolonged high T_A, hyperthermia and dehydration is somewhat better than that of birds from more moderate climates. Some species, at least, also appear to have an overall higher conductance than mesic birds. Desert birds do not have an overall higher T_B than mesic birds, though some do seem to have a lower RMR, which is an economy measure preventing excessive loss of water. The relatively thin plumage found in the Zebra Finch may facilitate heat loss from the body, but the thick plumage of sandgrouse, Ostriches and some waders may equally prevent heat gain from the environment; sandgrouse even have the front of the tarsometatarsus feathered. Birds have the advantage of a large respiratory surface area from which cooling may occur by evaporation; this happens also through the skin to a far greater extent than formerly recognized. The most significant conclusion is that birds are thermally preadapted to the arid zone by virtue of their relatively high T_B and the capacity to lower the RMR. Otherwise they simply use the basic avian equipment in order to withstand any rigours of their chosen habitats. Their behavioural thermoregulatory mechanisms are the same as those employed by birds in all habitats.

7 Breeding (1): Timing of Breeding

7.1 Introduction

With regard to the timing of breeding in birds, "by far the most important ultimate factor for nearly all species of birds is the availability of an adequate food supply. . . . Each species has therefore evolved the timing of its breeding so that it coincides with a maximum availability of its species-specific food" (Immelmann 1971). An exception to this may be the Double-banded Courser *Smutsornis africanus* in southern Africa, since it breeds throughout the year, regardless of environmental conditions (Maclean 1967b). "Every species is apparently adjusted to breed at the time of year at which it can raise its young most efficiently" (Immelmann 1971).

In this chapter the regulation of breeding will be dealt with according to geographical region, in order to highlight some of the differences which prevail in some of the deserts of the world. The regions covered are South America, Australia, Africa and North America. Within each region, certain species or groups of species have been selected for particular mention, since they illustrate special points of breeding ecology, the emphasis being on the initiation of breeding and the factors governing the duration of breeding seasons.

In the account which follows, the terms "proximate factor" and "ultimate factor" will be used in the sense of Immelmann (1972). Both apply to temporally and/or spatially restricted processes "for which a difference between selecting and regulating mechanisms is to be found". A proximate factor is an environmental stimulus which initiates or maintains biological processes (in particular breeding activity, but also moult, hibernation and so on) under the most favourable ecological conditions. An ultimate factor is an environmental factor which, through natural selection, has led to the relevant restriction of the processes to particular times within natural events; an ultimate factor has survival value. A proximate factor is merely one which initiates an activity at an appropriate time, so that the ultimate factors also operate appropriately later on. A *Zeitgeber* is merely a special kind of proximate factor of the same type as a releaser.

For example, birds of central Australia breed at any time of the year according to the unpredictable rainfall, which generates a short-lived but

Fig. 7.1. Black-faced Woodswallow *Artamus melanops*, Western Australia. (Photo by Graeme Chapman)

abundant supply of food (Immelmann 1963a). Rain itself may be a proximate factor (*Zeitgeber*) initiating breeding activity: indeed the sight of falling rain may be sufficient stimulus to release breeding in the Zebra Finch *Poephila guttata* and the Black-faced Woodswallow *Artamus melanops* (Fig. 7.1). Constant gonadotropic activity of the hypothalamus, the absence of a juvenile gonadal refractory period (a dormant phase before the initiation of gonadal activity at the start of a breeding season), the ability to moult and breed simultaneously, and factors facilitating pair formation contribute to the rapid response to rainfall seen in these arid-zone birds. The ultimate factor is food supply (Immelmann 1963b; Nix 1976). Primary *Zeitgeber* provide stimuli which initiate gonadal development, while secondary *Zeitgeber* operate at a later stage of gonadal development and maintain the congruence between the internal rhythm and the environment.

The internal cycle starts with a refractory period after breeding, during which the gonads undergo undisturbed regeneration (Immelmann 1963b). This is followed by a phase of reproductive readiness, during which tonic gonadotropic activity ensures that the animal is ready to reproduce at any time, given the correct set of environmental factors (*Zeitgeber*).

The internal cycle is linked to the environment through the *Zeitgeber* as defined above. *Zeitgeber* may be visual, auditory, behavioural and physiological (such as body condition). They can influence the annual cycle only between the end of the regeneration phase and the peak of development of the gonads. The more pronounced the variability in the external environment, the closer has to be the coordination between the endogenous rhythm of the gonads and the environmental fluctuations. In such regions as arid zones where environmental variability is great, development of the gonads has to proceed as far as possible without the influence of external stimuli. When the external stimuli (*Zeitgeber*) are favourable for the initiation of breeding, only the final stages of gonadal development have to be released, thereby ensuring that breeding follows in a very short time. Under such conditions the refractory period has to be abbreviated so as to allow the bird to breed even if the next set of favourable *Zeitgeber* follows shortly after the previous breeding season.

Recovery of populations of desert birds after droughts is the result of: (1) birds concentrating in environmentally favourable areas (nomadism); (2) protracted breeding during which several broods are reared; and (3) increased clutch size. These patterns are reflected repeatedly in the deserts of the world, notably those in the southern hemisphere and in the Gondwana provinces (including those parts of the Neotropical and Afrotropical regions which lie north of the equator) where climatic conditions are less predictable than in the northern hemisphere.

With decreasing latitude, seasonal differences are marked less by temperature than by alternating dry and wet periods. In the Spotted Munia *Lonchura punctulata* of India there are indications that, apart from decreasing photoperiod (the species breeds in autumn), food may be a proximate factor influencing breeding (Chandola et al. 1983). By abolishing the seasonal variation of food in captive conditions, the testicular cycle was shifted by 2 months every year, much as it was in birds maintained in constant darkness, when their gonadal cycle "free-ran" with a peak every 10 months.

7.2 South America

The peak of breeding in the birds of the Santa Elena Peninsula in arid southwestern Ecuador may vary by as much as 7 weeks from year to year as a result of varying times of rainfall (Marchant 1959). Similarly the duration of the breeding season varies from as little as 6 weeks in a year of poor rainfall to as much as 3.5 months in a year of good rainfall. In poor years certain pairs of birds may rear no broods because the young starve to death, but in good years up to three broods may be reared successfully. In contrast, the breeding seasons of desert and non-desert birds in Iraq are seasonal (Marchant 1963); seedeaters, such as sandgrouse, breed later than other species.

7.3 Australia

In the arid centre of Australia (Fig. 1.6) birds breed after rain, wherever and whenever it may fall (Keast 1959). Analysis shows that the breeding and seasonal movements of birds in the Australian region are closely correlated with seasonal pulses in plant growth, associated in turn with weather systems which determine the patterns of drought and rainfall (Nix 1976). Some species of birds react very quickly to rainfall: the Desert Chat *Ashbyia lovensis* (Fig. 7.2) may begin nestbuilding 5 days after the end of a drought and may ovulate on the 11th day. In the "intermediate" zones, birds undergo a spring gonadal recrudescence (enlargement of the gonads at the onset of reproductive activity) and then breed if environmental conditions permit. In a good season breeding may continue from spring to autumn, several broods being attempted in this time.

During the 1952 drought in Australia, the gonads of birds in arid regions (e.g. Alexandria) were fully regressed. Not far away at Camooweal after 25 mm of rain, local conditions were temporarily improved; males of several species began to produce spermatozoa, and the testes of all species tested were enlarged, but this was accompanied by a wide range of developmental abnormalities. No recrudescence of female reproductive organs took place at all. The inference is that the small rainfall had stimulated some testicular development, but in the absence of an adequate follow-up rainfall gonadal regression began to occur before any breeding took place. The environmental factors influencing breeding must be a combination of intrinsic and extrinsic factors.

Fig. 7.2. Desert Chat *Ashbyia lovensis*, New South Wales, Australia

7.3.1 The Zebra Finch *Poephila guttata*

The gonads of young Zebra Finches become mature at the age of about 10 weeks when they have almost completely attained adult plumage (Sossinka 1974, 1974/75, 1975). There is an interruption in testicular growth between the 18th and 34th days, at a stage when birds of more mesic environments undergo a gonadal refractory period of several months. This brief refractory period is, however, soon followed by further testicular development, a feature adaptive in an environment in which the next favourable breeding season may not occur for 2-3 years. So the young of a given breeding season have the potential to breed themselves in that same season, provided propitious conditions last that long.

Experimental changes in daylength and a 2-month period of water deprivation had no effect on the gonads of captive birds, though their thyroid glands showed increased activity.

Breeding by the Zebra Finch over a 3-year period in Western Australia did not necessarily follow rainfall in winter or in autumn (March; Davies 1977). Breeding rather occurred each spring, whether or not heavy rain had fallen, and in autumn only if rain had fallen in the previous summer. If T_A fell rapidly in autumn after rain, breeding did not occur. Zebra Finches feed on seeds of annual grasses; if seeding was poor, the birds did not put on sufficient weight to breed successfully. Breeding was therefore initiated in warm periods when T_A was rising, especially in spring, in years of effective summer rain when the grass had germinated and set substantial amounts of seed to serve as the birds' food supply. More young appeared to hatch in spring than in autumn or in winter. It is suggested (Davies 1977) that autumn breeding may occur in those years when autumn conditions mimic spring conditions.

7.3.2 Columbids

The breeding cycles of four Australian columbids (Crested Pigeon *Ocyphaps lophotes*, Common Bronzewing *Phaps chalcoptera*, Plumed (or Spinifex) Pigeon *Geophaps plumifera* and Redplumed Pigeon *G. ferruginea*) were studied in arid Australian habitats (Frith et al. 1976). Some males of all species had active gonads in all months of the year; fewer birds showed active spermatogenesis in autumn and winter than at other times of the year. The size of the testes did not reflect the stage of spermatogenesis, but the size of the testes showed an annual cyclical variation, being smallest in winter and largest in spring and/or summer when spermatogenesis was most active.

The effects of rainfall on testicular cycles was significant on all but the Common Bronzewing, but at no time did recent rain have a greater effect on the mass of the testes than did the testicular cyclic component when at its maximum in the annual cycle. Rain is particularly effective on testis activity of the Red-plumed Pigeon. The effect of rainfall was less marked on the birds of the semi-arid zone than on those of the arid zone where food supply is more erratic and more closely controlled by the timing and abundance of rain. The Plumed Pigeons continued to produce sperm and to lay eggs even in severe drought, except when heavy overgrazing depleted the seed supply; under natural conditions, therefore, a drought need not put an end to breeding in these birds (though the success of dry-season breeding has not been measured). Even in the arid environment, photoperiodic effects on the gonadal cycle are appropriate, in order to ensure increased readiness to breed in spring and early summer when even a light rain would result in an increased food supply.

The Plumed Pigeon shows an annual gonadal cycle, indicating that it is under the control of seasonal changes in day length, but it can also be strongly affected by rainfall which can modify the usual circannual cycle (Frith 1982). In times of drought, sexual development is greatest in drought-free districts. The adverse effects of drought on breeding condition are exacerbated by overgrazing by domestic stock which destroys the pigeons' food supply. Even so, the birds are still capable of breeding, unless the drought is really prolonged, such as happened at Windich Springs in the Gibson Desert, where no rain had fallen for 10 years. "By remaining sexually active in very dry weather it is in a suitable condition to take rapid advantage of any rain and breed" (Frith 1982). Although breeding may be attempted even in a drought, it is not known how successful such attempts are.

Fig. 7.3. Inland Dotterel *Peltohyas australis* adult in eclipse plumage

7.3.3 The Inland Dotterel *Peltohyas australis*

The Inland (or Australian) Dotterel (Fig. 3.1) has an eclipse plumage (Fig. 7.3) and therefore appears to have a set breeding season, which is unusual in birds of arid inland Australia (Maclean 1976b). From September 1974 to January 1975 the adults increasingly acquired eclipse plumage, even though many other bird species were breeding at the time as a result of a combination of good rains and perhaps also of photoperiodic influences. However, the synchrony of the dotterels' gonadal regression and subsequent recrudescence may have been a function of their having been brought into breeding condition at the same time by the previous rains, which resulted in their undergoing gonadal regression after the previous, rain-induced breeding season (though there was no evidence of young birds in the flocks); in the absence of information about the most recent breeding activity, this must remain speculative. Information about the timing of reproduction in this species is inconsistent (Maclean 1973a), but seems to point to nesting mainly between April and October, in line with the finding that the birds had completed breeding by September (Maclean 1976b). The Royal Australasian Ornithologists Union Nest Record Cards show breeding from February to October with peaks in February and June, again consistent with the finding that gonadal recrudescence usually begins in January (Maclean 1976b).

7.3.4 The Budgerigar *Melopsittacus undulatus*

The Budgerigar occurs as a breeding species throughout inland Australia where conditions vary greatly from south to north and from east to west. Movements and breeding are seasonal, following broad patterns of seasonal growth of pastures (Wyndham 1983). In the winter-rainfall southern inland the birds arrive in spring (September), breed, and leave in summer. In the mid-latitudes, residence and breeding are mostly in spring in the winter-rainfall mid-south, to autumn in the summer-rainfall mid-north. In the northern latitudes, Budgerigars are most abundant and breed in autumn and winter after the wet season (i.e. in the early dry season); numbers drop during the later dry season and during the rains. Similar patterns of movement and breeding schedules, involving movement southward in spring and northward in autumn, may exist also in other species of Australian birds, such as woodswallows *Artamus*, chats *Ephthianura*, trillers *Lalage*, the Inland Dotterel and the Australian Pratincole *Stiltia isabella*.

If rainfall is unseasonal, the movements of the Budgerigars may be affected; for example, after unusually abundant rainfall some birds may remain in an area during the winter when they would normally have departed. The conclusion is that movements and breeding are directly controlled by the abundance of food.

In a group of 41 captive-reared male Budgerigars, mature spermatozoa were first noted at the age of 3.5-4 months (Pohl-Apel and Sossinka 1975). No general response of the gonads to changes in photoperiod could be shown. (Perhaps domestication could have had an influence here, masking the response to photoperiod.)

7.4 Africa

7.4.1 Breeding Seasons

In semi-arid Senegal, with its marked wet/dry cycle, the rains average 320 mm a year and fall in the period July to October. Of the 213 breeding species of birds, 47% nested in the rains only, 34% in the dry season and 19% at any time of the year (M.Y. Morel 1973). Granivorous birds were much more numerous in terms of individuals than were insectivorous birds. They belonged to five families of tropical origin: Pteroclidae (one species), Columbidae (six species), Ploceidae (15 species) and Estrildidae and Viduidae combined (11 species). The length of the breeding season was either short and confined to the rains (Ploceidae, in which both adults and young fed to a large extent on insects during the breeding season), or long and continued into the dry season (the others, which fed essentially on seeds from hatching).

Adaptations to the semi-arid conditions in these granivores included the following (M.Y. Morel 1973):

1. The breeding season is always prolonged. Sandgrouse nested from the middle of the dry season into the rains (February to August or September). Doves nested from the end of the rains (September to November) and in the middle of the dry season (February to May), but sometimes all through the year. Estrildids bred from the start of the rains well into the dry season (July to May). It must be concluded that the *Zeitgeber* for the initiation of breeding was not the same for all these species.
2. The age of first breeding was accelerated so that birds could breed within the season in which they were hatched. Sandgrouse hatched at the start of the season could breed by the end of the same season. Doves hatched between September and November could breed between the following February and May: those hatched in the dry season could therefore breed as early as the following wet season.
3. The breeding plumage of such birds must follow the juvenile plumage immediately and rapidly.
4. The speed of the post-juvenile moult and the age at which it occurred were subject to environmental pressures. In the doves this moult started shortly after leaving the nest and was completed by the age of 6 months, so that the adult plumage was acquired at a very young age. Some individuals might in fact begin to breed before this moult was complete. In the estrildids, those hatched late in the season underwent an accelerated moult, so that all individuals hatched that season were in adult plumage at the start of the next breeding season, in spite of differences in age.
5. The skeleton of those birds breeding "precocially" was still in the process of pneumatization.

6. The population turnover was rapid. The annual adult mortality of the Red-billed Firefinch *Lagonosticta senegala* (Fig. 7.4) was about 70%. It was estimated that the mortality of doves was of the same order of magnitude, to judge by the small numbers of birds with full pneumatization.
7. In order to compensate for the high mortality, the reproductive rate must be increased. This was done in two ways: (a) by lowering the age at which sexual maturity was attained and (b) by increasing the duration of the breeding season in order to be able to attempt more broods. The Red-billed Firefinch might raise 4-5 broods in a season; it was estimated that each female produced an annual average of 12 eggs and raised an annual average of 3.3 young.

These reproductive strategies can evolve only in birds of small body size: they have relatively small energy requirements and make small demands on an environment of low productivity. Their rapid reproductive rate allows for accelerated genetic renewal and enhances the chances of new genetic combinations: this is reflected by the overall large numbers of species of Columbidae (255) and Estrildidae (108), which have diversified into a large number of microhabitats. They are also able to respond quickly to changes in the environment because of the flexibility of their mode of reproduction.

A lack of seasonal rain in the Sahara results in smaller clutches and fewer broods (usually only one) compared to good seasons (Casselton 1984). In drought years the gonads of birds may remain regressed and breeding will not occur. Some species may breed twice a year, depending on the food

Fig. 7.4. Red-billed Firefinch *Lagonosticta senegala*, South Africa. (Photo by G.K. Lestrange)

supply: granivorous species such as the Striped (or House) Bunting *Emberiza striolata* and doves breed in the spring when the barley is ripening, and in the autumn when the millet is harvested. The Cream-coloured Courser *Cursorius cursor* breeds throughout its range (the Canary Islands, North Africa, the Middle East, southern Russia and northwestern India) from February to June, mostly from April to May. The timing of rain may affect the timing of breeding, but only within the limits of the normal breeding season (Cramp and Reynolds 1972).

In the Kalahari, South Africa (Fig. 1.3), the Sociable Weaver *Philetairus socius* (Fig. 3.5) responds only to rain for breeding (Figs. 7.5 and 7.6), has the shortest recorded lag period of any bird (6 days) between the first good rain of the season and the first egg laid, and will continue to breed for as long as favourable conditions last, sometimes for up to 9 months, rearing four consecutive broods in the process (Maclean 1973d). Nest records show that the Namaqua Sandgrouse may nest in all months of the year (except March and May) in southern Africa, but most breeding is in the winter (May to August; Dixon and Louw 1978). Rainfall appears not to be a primary (proximate?) environmental cue for sexual activity, but rain and the accompanying cooler conditions in summer may initiate some "out-of-season" breeding.

In India and Pakistan the Chestnut-bellied Sandgrouse *Pterocles exustus* breeds the year after hatching (Christensen et al. 1964). It may rear two broods a year, one in each of the two major nesting periods, one from February to April and one from November to December (or January). Much the same seems to apply to the Black-bellied Sandgrouse *P. orientalis.*

Rainfall governs the breeding of all species of larks in the Kalahari sandveld (Maclean 1970a). The greater the amount of rain, the more species bred (with the exception of Stark's Lark *Eremalauda starki* (Fig. 5.3) after good summer rains, though populations at the time were low). Rainfall also governs populations of nomadic species to a large extent, but fluctuations of nomads at one site may also depend on rainfall elsewhere within the range of the species. However, the mere arrival of a nomadic species after rain is not necessarily accompanied by breeding in that species (such as in February and March 1965 when very large numbers of Black-eared Finchlarks *Eremopterix australis* (Fig. 7.7) and Stark's Larks moved into the Kalahari Gemsbok National Park after good rains, but did not breed there).

In line with most other species of small passerine, the Rufous-eared Warbler *Malcorus pectoralis* (Fig. 7.8) in the Kalahari sandveld breeds at any time of the year after rain (Maclean 1974a). It has a minimum lag period of only 7 days between the first good fall of the rainy season and the first egg. This response is only one day longer than that of the Sociable Weaver (Maclean 1973d).

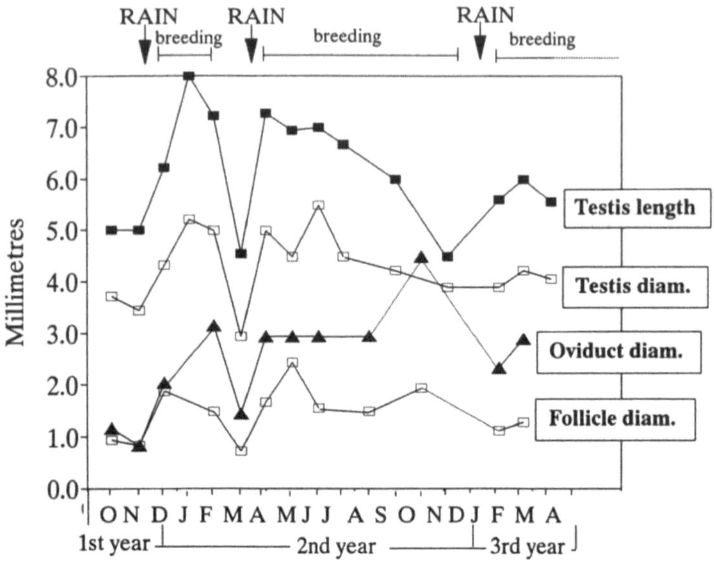

Fig. 7.5. Changes in size of reproductive organs of the Sociable Weaver *Philetairus socius* relative to rainfall in the Kalahari Desert, South Africa. (Redrawn from Maclean 1973d)

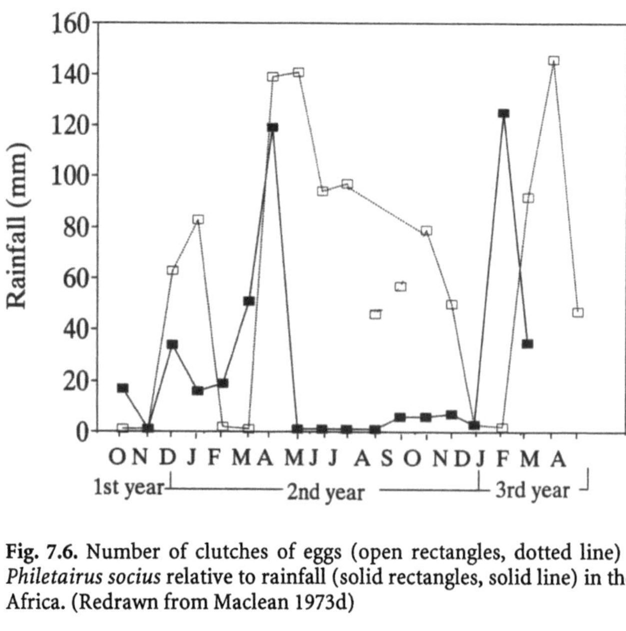

Fig. 7.6. Number of clutches of eggs (open rectangles, dotted line) of the Sociable Weaver *Philetairus socius* relative to rainfall (solid rectangles, solid line) in the Kalahari Desert, South Africa. (Redrawn from Maclean 1973d)

At the start of a breeding attempt the protein reserves of adult Red-billed Queleas *Quelea quelea* were higher than at other times of the year, especially in females (non-fat dry mass of flight muscle) in which an 80% increase in actual protein reserve was measured, as opposed to only 14% in males (Jones and Ward 1976). It is suggested that the state of the birds' protein reserves is the proximate factor controlling breeding, and that the birds need no environmental releasers to breed at the appropriate time of year.

In four species of hornbills in the genus *Tockus* in the Kruger National Park, South Africa (Yellow-billed Hornbill *T. flavirostris*, Red-billed Hornbill *T. erythrorhynchus* and Grey Hornbill *T. nasutus*: mean annual rainfall 570 mm), and in Namibia (Monteiro's Hornbill *T. monteiri*, Yellow-billed Hornbill and Grey Hornbill: mean annual rainfall 100 mm) the onset of nesting was linked to rainfall: no instances were recorded of nests started before the first rains of the wet season (Kemp and Kemp 1972; Kemp 1973). The proximate factors initiating breeding are not easy to determine, but food supply is an especially important limiting factor during the breeding season. Rainfall not only brings on an increased supply of arthropod food, but it also makes mud available for sealing of the nest hole (except in the Grey Hornbill which does not use mud). However, mud is not believed to be an essential proximate factor in the initiation of breeding, since pairs nesting near an artificially provided source of mud do not nest earlier than pairs nesting under completely natural conditions. The conclusion is that food is the proximate factor. This is borne out by the fact that the arboreally foraging Grey Hornbill nests up to a month earlier than the other species which forage on the ground: the Grey Hornbill forages in deciduous trees which come into leaf well before the rains, and probably provide the birds with increased arthropod food even before the rains start.

Hornbills in the drier Namibian environment have a lower threshold of response to rain than those in the wetter lowland habitat of the Kruger National Park, probably because the birds need to nest as soon as possible after rain, and also perhaps because the response of the arthropods to rain may be more marked in the drier region. The lag period between the first heavy rains of the wet season and the laying of the first egg was 23-35 days. In 15 nests the first eggs were laid within a period of only 12 days.

The correlation between rainfall, availability of fruit and the breeding in the White-backed Mousebird *Colius colius* (Fig. 4.10) in the semi-arid Karoo of South Africa is weak (Dean et al. 1993). The species breeds opportunistically at any time of the year, but not during the coldest months because of its intrinsic thermoregulatory problems (it sunbathes at any time of the day to build up heat). It is suggested that the availability of new leaves and flower buds may provide a cue to the initiation of breeding in this species.

Fig. 7.7. Black-eared Finchlark *Eremopterix australis* male at nest, Kalahari Desert, Botswana

Fig. 7.8. Rufous-eared Warbler *Malcorus pectoralis*, Kalahari Desert, Botswana

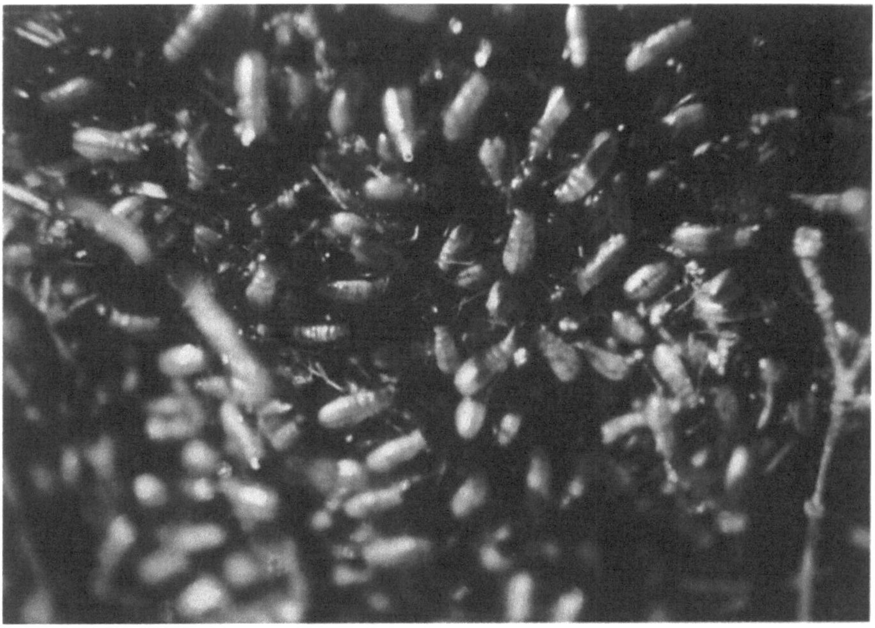

Fig. 7.9. Harvester Termites *Hodotermes mossambicus*, a widespread and abundant source of food and water for birds in arid southern Africa

7.5 Africa and Australia Compared

A comparison between arid-zone birds of the Kalahari and Australia (Immelmann 1967, 1970) showed that the African birds had a more obvious internal rhythm of reproduction, were much less dependent on a single external *Zeitgeber*, and therefore exhibited greater individual variability in the timing of their breeding. This might be the result of a number of factors:

1. In Australia breeding can occur only after rain because only then is there adequate food for rearing young.
2. The sandy soils of the Kalahari hold water for long periods, so that the effects of rain on the vegetation may be less dramatic.
3. The almost daily availability of the diurnal Harvester Termite *Hodotermes mossambicus* (Fig. 7.9) ensures a more evenly distributed supply of insect food throughout the year in the Kalahari.

However, Immelmann's (1967) study was based on a single season's observations in the Kalahari and therefore is heavily influenced by the events of that season. A more prolonged study would have shown that, in certain seasons at any rate, the dependence of the birds on rainfall was as great as that in the birds of the Australian arid zone (Maclean 1970b).

Fig. 7.10. Desert-scrub habitat in Arizona, USA

7.6 North America

About 56 species of birds commonly breed in the desert-scrub habitat of the Sonoran Desert [southwestern Arizona (Fig. 7.10), northwestern Mexico and Baja California; Vleck 1993]. Five species nest mainly in summer, 25 mainly in spring and 19 in both spring and summer, so that the breeding season may be long, from 6-12 months. Sixteen of the species listed for the Sonoran Desert have nesting seasons of 6 months or longer, probably because temperatures are seldom low enough to inhibit breeding. Multiple broods are common, up to four in the Curve-billed Thrasher *Toxostoma curvirostre* and the Cactus Wren *Campylorhynchus brunneicapillus*, though two to three broods are more usual.

7.6.1 The Cactus Wren *Campylorhynchus brunneicapillus*

The Cactus Wren is a common resident of the desert of southwestern North America to central Mexico. The principal food of nestling wrens in the Chihuahuan Desert of New Mexico is the Band-winged Grasshopper *Trimerotropis*, forming 95% of the food delivered (Marr and Raitt 1983). Marked differences in the abundance of *Trimerotropis* affect the breeding ecology of Cactus Wrens and probably depend on differences in rainfall, since the rainfall affects the annual plants on which the grasshoppers feed.

This grasshopper overwinters in the soil and emerges in spring after the soil has warmed over several days. The abundance of its food plants depends on the amount of rain in the previous October and November (autumn and early winter).

The start of egglaying at the beginning of a breeding season in the Cactus Wren is associated with sharp rises in daytime temperatures over 3-4 consecutive days, which usually occurs in the last week of April. Parents feed nestlings in the early morning hours to avoid the greatest heat of the day; if early morning temperatures are below normal, feeding schedules are disrupted because the female has to spend more time brooding the young, hence the need to start breeding only when daytime temperatures are rising; this is also associated with the emergence of the grasshoppers and predicts favourable conditions at a time when nestlings are being fed.

Cactus Wrens reproduce at a level that incurs food limitation (Simons and Martin 1990). Food availability can limit reproduction in certain years; the greater the food supply the heavier the chicks and the greater their survival rate. Supplemental feeding in experimental birds can enhance survival rate in poorer years, and results in adults increasing their frequency of second broods by reducing the energy cost of producing the first brood. It appears that the clutch size has evolved to take advantage of good years when food is more abundant, a strategy that may compensate for the unpredictability of primary production in deserts.

Reproductivity may also depend on population densities, since food limitation of broods may be more intense in years of good food supply (than in years of poor food supply) when population densities are high, as more time may be spent defending territories or warding off competitive species such as the Curve-billed Thrasher. Ambient temperatures may also have an effect. The earlier the onset of high spring and summer temperatures (above 32 °C), the lower the activities of the wrens and the lower therefore the foraging activity.

7.6.2 The Roadrunner *Geococcyx californianus*

In the Tucson area of Arizona the Roadrunner (Fig. 5.6) has a bimodal breeding schedule, one peak being from mid-April to mid-June, the second from late July to mid-September (Ohmart 1973). The peaks are separated by a period of extreme heat and aridity in late June and early July. A similar bimodal breeding schedule is shown by several other species which feed their young on animal food, namely the Brown Towhee *Pipilo fuscus*, Rufous-winged Sparrow *Aimophila carpalis* and Curve-billed Thrasher, undoubtedly because food in the form of insects and other animal prey is most available during the two breeding peaks. By contrast, the granivorous American Mourning Dove *Zenaidura macroura* (Fig. 4.5) breeds continuously from January to October. Some of the young hatched early in a given season may themselves breed towards the end of the same season!

As in the Cactus Wren, clutch size in the Roadrunner may be influenced by food supply. The mean clutch size increases from spring to summer, since the Sonoran Desert is much more productive after the summer rains than in spring after a normal winter. Clutch size in the Roadrunner may vary from 2-12 eggs, though the largest clutches may be laid by more than one female. As environmental conditions deteriorate before the summer rains, clutch size is reduced to as few as two eggs, and young in the nest at such times are below normal body weight: a female which produced a clutch of two eggs in a lean period laid a clutch of six after the summer rains. In times of abundance of whiptailed lizards (*Cnemidophorus*) the young gain weight rapidly, but when these lizards are scarce and young have to be fed on a greater proportion of insects, they are underfed, especially when they are older than about 6-8 days. The young normally receive larger food items than are eaten by the adults.

7.6.3 Breeding Seasons

Breeding in some species is associated with rainfall. For example the Rufous-winged Sparrow usually breeds after the summer rains in July after a slow gonadal recrudescence from March to June; however, after heavy winter rains the gonadal recrudescence is rapid and the species breeds in spring as well as in summer. The Black-throated Sparrow *Amphispiza bilineata*, on the other hand, breeds in winter in Baja California because of winter rains; in Arizona with its bimodal rainy season, the species breeds in both spring and summer.

Environmental cues initiating gonadal recrudescence in winter in these species are not known, but are likely to involve daylength, because recrudescence always occurs during the weeks preceding actual breeding. Gonadal recrudescence in the Curve-billed Thrasher coincides with the winter solstice, so it may be under the influence of short days rather than increasing photoperiod, whereas in the Cactus Wren and Black-throated Sparrow, gonadal recrudescence begins with the start of increasing photoperiod.

Gonadal cycles of Sonoran birds are photoperiodic and therefore differ from the cycles of such Australian species as the Zebra Finch whose gonads may be continuously active. They are more similar to the cycles of the Budgerigar whose gonads are inactive when the birds are not breeding. Sonoran birds seldom breed in autumn because food resources are low. Environmental parameters are predictable in Sonora, so that breeding is seasonal and only slightly modified by local rainfall; rainfall, however, is closely correlated with population numbers (Vleck 1993).

Luteinizing hormone (LH) and testosterone levels in the blood of male birds in Sonora rise as gonadal recrudescence is occurring, but these concentrations tail off as the breeding season progresses, even though the gonads remain active and multiple broods are common. The highest levels of reproductive hormones in males are associated with the start of the breeding season when male–male interactions are strongest, but such levels are not necessary for the continued maintenance of secondary sexual characteristics or sexual behaviour. As the season progresses, male–male interactions become less intense as testosterone levels decline, but breeding continues.

Reproductive hormone levels in cooperatively breeding desert birds, such as Harris's Hawk *Parabuteo unicinctus*, are high in the alpha male and in unrelated male helpers, but are low in related male helpers. Unrelated helpers are reproductively competent and may occasionally copulate with the alpha female. Testosterone levels in all primary and secondary breeding males are, however, very low, allowing for a harmoniously cooperative group with enough testosterone to react aggressively to other groups of hawks.

7.7 Moult

The moult in most of 20 species of arid-zone birds studied in Australia lasted 4-4.5 months (Keast 1968). Migrants have shorter moulting periods than residents. The duration of moult is also directly related to body size: the smaller birds have shorter moults. Moulting is more regular than breeding, so that a "post-nuptial" moult will occur even if breeding has not taken place. Likewise, good rains will initiate breeding, even though the birds may be in full moult. The Sociable Weaver (Fig. 3.5) of the Kalahari desert is somewhat unusual in that it moults the body plumage within 1 month, usually after rain when the birds are breeding, but remex moult takes about 9 months, since each remex takes about 1 month for replacement (Maclean 1973d).

7.8 Conclusions

Most arid-zone birds respond to rainfall by coming into breeding condition, even in winter, especially in the more unpredictable deserts of the southern hemisphere. Nomadic species move into areas where rain has recently fallen, but may not breed there at once or at all (possibly because they have recently bred elsewhere). Proximate factors that govern breeding in arid-zone birds include photoperiod (increasing or decreasing), rainfall and food supply; temperature may also be important in that it accelerates emergences of insects. A certain minimal amount of rain must have fallen to induce breeding, probably not less than about 30 mm as a rule. The amount of rain determines the degree of breeding response in birds, in terms of the length of the lag period (the time between the first good rains and the first eggs of the season), clutch size and breeding success. The lag period in many species of desert birds is shortened by their having tonic gonadotropic activity, by which the gonads come into a state of enhanced recrudescence in the absence of environmental stimuli; this applies especially to birds in the deserts of the southern hemisphere.

Breeding in most small arid-zone birds continues as long as favourable conditions last; this may be up to 9 months and involve the raising of up to four consecutive broods. The ultimate factor in avian breeding ecology is always food supply, regardless of the habitat (Nix 1976). Moult and breeding may occur simultaneously, but moult is usually slowed down if breeding is initiated while moult is in progress. Seedeating birds usually start to breed later than insectivorous species. The age of first breeding in some desert birds is reduced to less than a year, so that some young (e.g. estrildids, doves) may breed in the same season as that in which they were raised.

It is clear from this review that most work on the initiation of breeding in desert birds (and, indeed, most work of any kind on desert birds) has been done in southern Africa, North Africa, North America and Australia. The arid regions of South America, the Middle and Far East, and the Horn of Africa have been much neglected, opening up great possibilities for future research on desert birds. With increasing desertification of the borders of the world's true deserts, this kind of research is becoming increasingly important.

8 Breeding (2): Ecology of Breeding

Once breeding activity has been initiated under the influence of the various environmental stimuli described in the previous chapter, courtship, nestbuilding (where relevant) and subsequent parental behaviour follow in sequence. These activities may involve finding appropriate nesting materials (for example, green grass in the case of weavers *Ploceus* spp.), acquiring enough food to build up body condition and reserves for the production of eggs, energy to meet the behavioural demands placed upon the birds, and so on. Because of the extremes of the arid environment, factors such as nest orientation (mainly for shade in the heat of the day), and parental protection of eggs and young against temperature fluctuations and predation, are important for the birds' survival. This chapter will examine the ecology of breeding in several groups of birds, with the accent on parental behaviour and nesting adaptations, and will end with a set of conclusions, embodying the general principles to be derived from the results of the studies examined.

8.1 Sandgrouse

Namaqua Sandgrouse *Pterocles namaqua* (Fig. 3.8) nest on the ground (Fig. 8.1), sometimes exposed, at other times next to or under the shelter of a shrub or tuft of grass. The male incubates at night, the female by day after she has been to drink at the watering place (Maclean 1968). After being relieved in the evening, the female may again fly to drink. Neither parent leaves the nest during its attentive period, unless disturbed. In summer the male incubates for about 14 h, the female for about 10 h; in winter the male incubates for about 16 h and the female for about 8 h. Incubation in both Pin-tailed Sandgrouse *Pterocles alchata* and Spotted Sandgrouse *P. senegallus* is also continuous, the male incubating from about 18:00 to about 08:00 h in Pin-tailed Sandgrouse, and from about 19:00 to about 09:30 h in Spotted Sandgrouse, the females for the remaining daylight hours (Marchant 1961). The eggs are never left unattended unless the incubating bird is disturbed from the nest. Survival of clutches in Iraq is extremely low because of trampling by domestic stock and predation by unknown agencies (possibly foxes or ravens).

Fig. 8.1. Nest and eggs of Namaqua Sandgrouse *Pterocles namaqua*, Namib Desert, Namibia

Incubating Namaqua Sandgrouse in the Namib Desert raise the body from the eggs under high T_A, presumably to allow convective cooling of the eggs, and face into the sun, presumably to minimize the surface area exposed to direct radiation; but if a cool wind is blowing the birds face into the wind (Dixon and Louw 1978). Since the eggs rest on the ground (Fig. 8.1), soil-surface temperatures are a critical factor in the breeding success of sandgrouse. When soil temperatures approach 60 °C the incubating bird is unable to regulate the temperature of the nest effectively, so that it may exceed 45 °C, despite intense thermoregulatory behaviour by the incubating bird. Under such conditions the embryos do not survive, which may be why this species nests mainly in the winter months when soil temperatures do not exceed 49.3 °C at 14:00 h. By contrast the summer (October to April) soil temperatures at 14:00 h can range from 54.3 °C to 58.9 °C. If soil temperatures do not exceed 50 °C, however, the incubating sandgrouse can keep the temperature at the surface of the eggs below 40 °C (Fig. 8.2).

Namaqua Sandgrouse eggs are smaller (about 11.1 g freshly laid, which is only about 43% of expected) than might be expected from a typical charadriiform of equivalent body mass (about 25.9 g freshly laid), and the incubation period is rather shorter, but corresponds approximately to that for an egg of its size (Thomas and Maclean 1981). This apparent reduction in the size of the egg does not apply to other species of sandgrouse studied (Hinsley and Ferns 1994) and may therefore not have the survival value

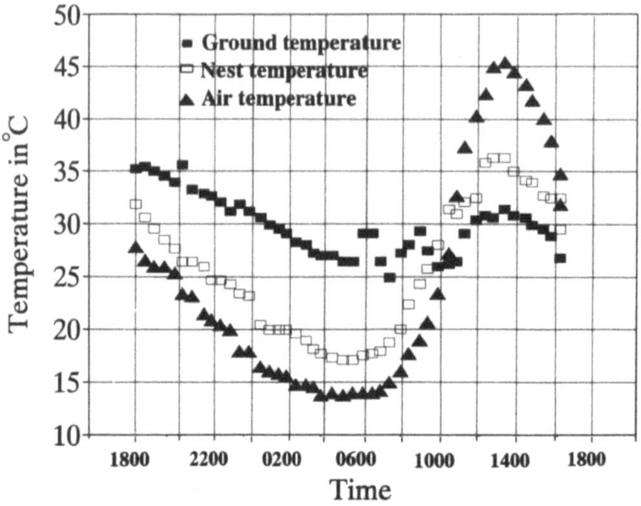

Fig. 8.2. Ground, air and nest temperatures at nest of Namaqua Sandgrouse *Pterocles namaqua*, Namib Desert, Namibia. (Redrawn from Dixon and Louw 1978)

ascribed to it by Thomas (1984c). Namaqua Sandgrouse shade their eggs in the heat of the day (unlike species in the northern hemisphere: Pintailed Sandgrouse, Crowned Sandgrouse *P. coronatus* and Spotted Sandgrouse), so the eggs would tend to lose water vapour to the surrounding air more easily than if the bird's body were in contact with them. However, the eggs are in fact adapted to reduce water-vapour conductance, despite a relatively thin eggshell (Thomas and Maclean 1981). This decrease in conductance is partly ascribable to a disproportionate decrease in total functional pore area. Two advantages of having relatively small eggs in a hot, open habitat with low productivity would be that (1) the clutch can be formed with relatively low investment of energy; and (2) the shorter incubation period reduces the time during which the clutch is vulnerable to predators or temperature extremes.

After hatching, sandgrouse chicks feed by themselves on seeds, and the male parent brings them water to drink in his belly feathers (Meade-Waldo 1896; Marchant 1961, 1962; Cade and Maclean 1967; Maclean 1983).Writing about Pin-tailed Sandgrouse nesting in captivity Meade-Waldo (1896) stated: "As soon as the young were out of the nest (when twelve hours old) a very curious habit developed itself in the male. He would rub his breast violently up and down on the ground, a motion quite distinct from dusting, and when all awry he would get into his drinking water and saturate the feathers of the under parts. When soaked he would go through the motions of flying away, nodding his head, &c. Then, remembering his family were close by, would run up to the hen, make a demonstration, when the young would run out, get under him, and suck the water from his breast. This is no doubt the way that water is conveyed to the young when hatched far out on waterless plains. The young, which are most beautiful little creatures, and very difficult to see even in an aviary, are very independent, eating hard seeds and weeds from the first, and roosting independently of their parents at ten days old."

The first field observations of this behaviour were of a male Spotted Sandgrouse in Iraq supplying the chicks with water from his belly plumage (Marchant 1961). Three different observations of young drinking from the belly feathers of male Pin-tailed Sandgrouse confirm that this is the usual method of watering (Marchant 1962). Marchant calls it "litter-of-puppies" behaviour. Even young already able to fly obtain their water in this way.

Cade and Maclean (1967) and George (1969) confirmed that watering of the young in the Namaqua and Spotted Sandgrouse (and usually in others too) is done solely by the male, and starts immediately after the young hatch. Only the central belly feathers take up water; the lateral feathers cover about 30% of the wet central plumage during the return flight from the waterhole. Chicks react only to the call of their own parents, their vocal response indicating to the parents where the young are located. On arrival the male stops about 1-2 m from the young; adopts an upright posture and spreads the belly plumage; this immediately signals the chicks to come running to

Fig. 8.3. Namaqua Sandgrouse *Pterocles namaqua* male with chicks drinking from his soaked belly plumage (only one chick visible), Kalahari Desert, Botswana

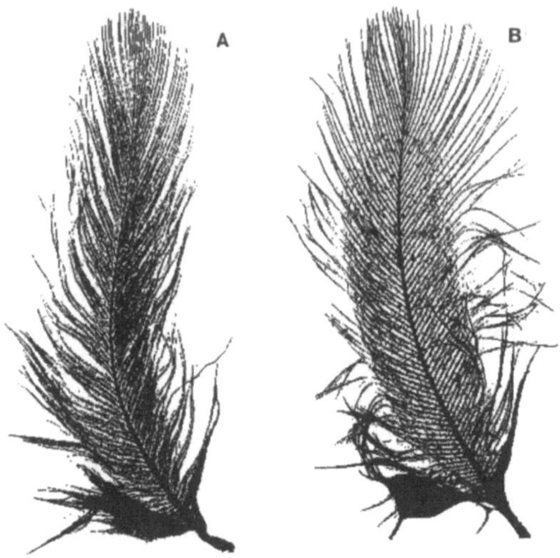

Fig. 8.4. Namaqua Sandgrouse *Pterocles namaqua* belly feathers (*A*) dry and (*B*) wet.

him and take water from his belly plumage, sometimes for as long as 10 min (Fig. 8.3). After watering the young, the male walks to a sandy patch where he sandbathes to remove the remaining moisture in his plumage.

In their detailed study of this remarkable parental behaviour Cade and Maclean (1967) and (Maclean 1983) observed male Namaqua Sandgrouse and Burchell's Sandgrouse *P. burchelli* in the Kalahari Desert of South Africa soaking their belly feathers in order to carry water to their chicks. First the male parent rubs his belly feathers in sandy soil, presumably to remove any waterproofing preen oil. He then walks into the water until it touches his belly plumage. He lowers his body into the water, spreads the ventral plumage and rocks the body up and down, pivoting on the synsacrum, in short quick bursts of five to six rocks at intervals of a few seconds to a minute or more. Tail and wings are held clear of the water. On his return from the watering place a belly-soaked male adopts an upright stance, calls briefly and the chicks rush to him to drink from the wet plumage with downward jerks of the head as if stripping water from the feathers. A drinking bout may take as long as 5 min, after which the male rubs his belly plumage in the sand as already stated, presumably to dry it.

Sandgrouse belly feathers are long and narrow with a slight curvature away from the midline of the body (Cade and Maclean 1967). The elongate shape may be adapted to being stripped of water by the chicks and the curvature of the feather could retard downward movement of the water; this shape also means that there is considerable overlap of the belly feathers, only the distal fifth or so being exposed when the plumage is intact. The proximal zone of the inner surface of each feather consists of specially modified barbules: the base of each barbule is somewhat flattened and spirally coiled in the dry condition, while the tip is a straight filament about 1 mm long. When dry, adjacent barbules intertwine (Fig. 8.4A), but on wetting the barbules uncoil (Fig. 8.4B), causing the distal filaments to stand vertically to the feather vane, forming a dense bed of fine hairs which hold the trapped water. The feathers of female sandgrouse are similar to those of males, but the area of specialized barbules is smaller. The males of all sandgrouse, except the Tibetan Sandgrouse *Syrrhaptes tibetanus*, have this specialized feather structure. The Tibetan Sandgrouse is an inhabitant of high, cold mountains where insulation is important and where water from snowmelt is always available: this sandgrouse does not fly to water as regularly as do other species; presumably the chicks drink water themselves or feed on succulent food.

The belly feathers of sandgrouse are superior to all other kinds of feathers tested in their capacity to hold water (Cade and Maclean 1967). The belly feathers of male Namaqua Sandgrouse, Burchell's Sandgrouse, Yellow-throated Sandgrouse *Pterocles gutturalis* and Double-banded Sandgrouse *P. bicinctus* typically hold 15-20 mg of water/mg dry weight of feather. The feathers of seven other species of birds of different families held only 5-6 mg water/mg dry weight. Furthermore sandgrouse feathers retain their structural integrity when repeatedly wetted and stripped of water with the fingers, while feathers of other birds tend to become frayed, matted and twisted under this treatment.

Histological, scanning-electron and X-ray-crystallographic studies of sandgrouse feather barbules show that the barbule is helically coiled for 3.5 turns and is kidney-shaped in transverse section with a pitted concave inner surface and a smooth convex outer surface (Joubert and Maclean 1973; Maclean 1983). The keratin is crystalline when dry, this crystallinity being somewhat reduced on wetting. When wet, the barbule uncoils as a result of expansion of the polypeptide chains of β-keratin in order to accommodate additional water bound to the side chains.

A single male Namaqua Sandgrouse can carry 25-40 ml of water in his belly plumage (Cade and Maclean 1967). Evaporation of the water in flight is reduced by the position of the water-carrying surface on the inside of the feather, by the overlap of adjacent belly feathers, and by flying to and from the watering place in the cooler morning hours (or at night in the case of the Doublebanded Sandgrouse and its relatives of the subgenus *Nyctiperdix*). It is calculated that a male Namaqua Sandgrouse starting off with 25 ml in his plumage can deliver 10-18 ml of water over a distance of 30 km. Small chicks of the Namaqua Sandgrouse (Fig. 2.4) in the Namib Desert may have up to 3 ml of water in their crops at 16:00 h, which is well within the theoretical calculations of Cade and Maclean (1967) (Dixon and Louw 1978). It is not known how much water a juvenile sandgrouse needs per day, but it is probably very little. Adult sandgrouse drink only 10-36 ml of water per visit to the drinking place or between 14 and 16% of body mass (Cade and Maclean 1967).

Chicks of the Pin-tailed Sandgrouse in captivity drank readily from the rubber of a dropper or from the space between the fingers of a moistened hand (Von Frisch 1968). They would not drink from standing water, no matter how thirsty they were. Young birds drank standing water only when able to fly. A translation of von Frisch's statements runs as follows:
"Some of our observations agree with the field observations of Cade and Maclean. We did not witness any watering of young in the Pintailed Sandgrouse in the field, and the great need for liquid in the young actually militates against this. The adults fly mainly at night to drink and not in the heat of the day when the chicks most needed fluids. Perhaps they are nevertheless given water or predigested food from the crop? I cannot say anything more precise about that yet."

Captive chicks never ate any animal food that was offered (Von Frisch 1969). Although von Frisch did not doubt the observations of Cade and Maclean (1967), his captive chicks required water at least every hour throughout the day, drinking as much as 20-25 g each up to five times a day. Unless the seeds offered as food were augmented with a dilute solution of oatmeal, the chicks did not appear to get enough nourishment. The differences between the observations of these different workers cannot immediately be explained, but may have something to do with the birds' behaviour in captivity.

Water transport in the belly feathers (*not* breast feathers, as stated by Thomas and Robin 1977) may be used to moisten eggs, as well as to water chicks, since a shot male Crowned Sandgrouse had wet ventral plumage after just having relieved the female on a clutch of fresh eggs (Thomas and Robin 1977). Experiments confirmed the studies of Cade and Maclean (1967) that sandgrouse feathers take up more water more quickly than do feathers of non-sandgrouse origin. Increased water-carrying capacity of sandgrouse feathers is directly related to their increasing adaptation to aridity in the following sequence: *P. bicinctus* < *P. burchelli* < *P. gutturalis* < *P. coronatus* < *P. namaqua* < *P. senegallus* < *P. lichtensteini*. Only the position of *P. lichtensteini* is anomalous in this sequence.

In other, though distantly related charadriiforms and ground-nesting caprimulgiforms, belly-soaking is known to occur for the purposes of either cooling chicks or wetting eggs (Maclean 1975; Grant 1982). For example, parent Little Ringed Plovers *Charadrius dubius* in Germany at T_A 29.2 °C wetted their belly plumage every 4-10 min by running through the water with the plumage spread, then ran or flew to the nest where they covered the chicks with the wet plumage (Gatter 1971). Whether or not the chicks also drank the water from the feathers could not be ascertained. The White-crowned Plover *Vanellus albiceps* of tropical Africa soaks its belly feathers to wet the eggs during incubation on hot days (Begg and Maclean 1976). These observations provide further evidence for a relationship between the Charadrii and Pterocli. Perhaps the main advantage of carrying water in the belly plumage, rather than in the crop, is that the parental sandgrouse does not need to sacrifice any of its own internal water supply for the chicks, even though belly-soaking involves carrying extra weight.

In the heat of the day, chicks 5-6 days old of the Chestnut-bellied Sandgrouse in Kaira District, India, "were constantly making an attempt to keep under the tails or next to the bodies of one of the parent birds" and appeared to suffer from the heat (Aldrich 1943). The same behaviour occurs in the Namaqua Sandgrouse (Maclean 1968).

Although moulting birds can be found at any time of the year (Dixon and Louw 1978), most individual Namaqua Sandgrouse in the Namib Desert moult from May to August (i.e. during the winter, when most of them are breeding). Moult in the Chestnut-bellied Sandgrouse and the Black-faced Sandgrouse *P. decoratus* in Tanzania seems to be even less tied to the time of year: Kalchreuter (1979) indicated that primary remex moult occurred over the whole year, and some primaries might be retained for two years. Body moult took place twice a year, but not in marked accordance with any season. However, moult in these two species was usually completed by the start of the extended breeding season in May. The timing of moult and breeding is a valuable field of further study, not only in sandgrouse, but in all desert birds, since both of these activities are expensive of energy, yet they may occur simultaneously, especially in the deserts of the southern hemisphere.

Sandgrouse must be among the most interesting of all desert birds to study because they are unusual in several respects, but, in the words of Meade-Waldo (1897): "I could write much more on these birds, but fear to trespass further."

8.2 Other Charadriiformes

Breeding in the Double-banded Courser *Smutsornis africanus* in the Kalahari (Fig. 2.3) occurs throughout the year, the number of nests in any area being directly related to the population of Double-banded Coursers present (Maclean 1967b). This points to the probability that the birds are multiple-brooded. The clutch is one egg. Of 56 eggs observed in this study none was known to be infertile and parental care of the egg is highly developed. Perhaps because of the small clutch size and the protection by the parents, predation on eggs is very low. A female may lay the egg of a second brood when the chick of the previous brood is still being fed. Having a single chick at a time allows a pair of coursers to rear a chick even in the driest of seasons. The combination of a single-egg clutch and sequential layings suggest the evolution of an "exploded" clutch of three to four eggs per year, the product of each laying being reared before the next egg is laid.

At T_A below about 21 °C the Double-banded Courser incubates the egg continuously (Fig. 2.3). Attendance at the nest become erratic between T_A 21 and 31 °C, but above that the parent is always in attendance, especially if T_A exceeds 35 °C on a sunny day. In this temperature range the egg is often shaded (Fig. 8.5), but at T_A above 36 °C the egg is incubated, the sitting bird keeping its legs exposed in order to lose heat by convection through these

Fig. 8.5. Double-banded Courser *Smutsornis africanus* shading exposed egg at T_A 35 °C

Fig. 8.6. Double-banded Courser *Smutsornis africanus* covering egg, but keeping legs shaded and exposed at T_A >38 °C

Fig. 8.7. Nest and eggs of Australian Pratincole *Stiltia isabella*, New South Wales, Australia

thermal windows (Fig. 8.6). Embryos die very quickly if unprotected at high T_A. Nest relief occurs every 2 h on average. The off-duty parent often stands for the entire off-duty period in the shade of a shrub without feeding, since feeding is usually done at dawn and dusk, or at night if the moon is bright.

The Cream-coloured Courser *Cursorius cursor* may leave its eggs unattended for long periods in dull weather, but in full sunshine it is reluctant to leave the eggs and returns within a few minutes (Cramp and Reynolds 1972). Under such conditions it crouches over the eggs to shade them, spreads the wings slightly, ruffles the dorsal plumage and keeps the exposed legs shaded. It pants for evaporative cooling.

The Australian Pratincole *Stiltia isabella* (Fig. 4.7), like other Charadrii, nests in the open on the desert floor (Fig. 8.7; Maclean 1976a). Both sexes incubate for a mean of 1.5 h at a time during the daylight hours. Attentive periods are shorter at higher T_A. Incubating birds pant and erect the dorsal plumage (Fig. 8.8); they do not gular flutter, nor do they ever merely shade the eggs. On sunny days incubation is continuous, but on cloudy days the eggs may be left unattended at T_A around 35 °C. Unlike most charadriiforms, however (for example the Snowy Plover *Charadrius alexandrinus* on the Great Salt Plains of Oklahoma, USA; Purdue 1976), the Australian Pratincole does not shade its eggs by squatting or standing over them in the heat of the day.

Fig. 8.8. Australian Pratincole *Stiltia isabella* incubating eggs at T_A >35 °C; raised dorsal plumage and panting are characteristic heat-loss mechanisms

On hatching, the chicks seek shelter under a nearby shrub or grasstuft. A parent arriving with food lands a few metres from the chicks' refuges and runs towards them; the chicks come out to be fed and return to the refuge. If undisturbed, the chicks crouch in the shade at the base of the plant, but creep deeper into the sheltering plant on disturbance. Parents lead chicks to a new refuge up to 100 m away if unduly disturbed.

Only in the Charadriiformes, among desert birds, does egg covering routinely take place, and then only in two unrelated groups: the seedsnipe (Thinocoridae) of South America (Maclean 1969, 1974c) and the Inland Dotterel *Peltohyas australis* of Australia (Maclean 1973a, 1974c). Maclean (1974c) has indicated that egg covering serves two functions: (1) concealment; and (2) insulation against loss of heat or against insolation. In view of either of these possible purposes, it is interesting that this behaviour has not arisen in more arid-zone birds, especially those in hot deserts.

8.3 Pigeons

Feral Pigeons *Columba livia* nesting under conditions of high T_A and low RH, when compared with conspecifics nesting in urban habitats and with pure wild stock nesting under mesic conditions, had eggs with an approximately

30-40% lower water-vapour conductance across the shell, significantly reduced effective pore area in the shell and a greater thickness of shell than was predicted (Arad et al. 1988). The differences are obviously adaptive to the environmental conditions under which the pigeons are breeding, but it would be useful to know if they are the result of acclimation or of genetic differences between the different populations of pigeon.

Acclimated experimental pigeons incubated their eggs and reared broods of young successfully at T_A between 45 and 50 °C, temperatures which are routinely experienced in the wild (Marder and Gavrieli-Levin 1986). Table 8.1 shows the temperatures of body and egg (T_{egg}) of incubating and non-incubating pigeons at different T_A, from which it is clear that incubating birds are capable of keeping both T_B and T_{egg} within viable limits, even at T_A of 50 °C. Indeed the incubating birds had lower T_B on average than non-incubating birds, lowering the T_B largely by means of CEWL.

Table 8.1. Mean temperatures (°C) of body (T_B) and egg (T_{egg}) of incubating and non-incubating Rock Doves *Columba livia* at different ambient temperatures (T_A). (Marder and Gavrieli-Levin 1986)

T_A	T_B non-incubating	Incubating	
		T_B	T_{egg}
28	41.0 ($n = 9$)	40.8 ($n = 9$)	36.7 ($n = 8$)
45	41.2 ($n = 5$)	38.9 ($n = 5$)	39.0 ($n = 8$)
50	42.4 ($n = 4$)	40.0 ($n = 4$)	40.7 ($n = 6$)

8.4 The Roadrunner *Geococcyx californianus* (Fig. 5.6)

Nests of the Roadrunner in the Tucson area, Arizona, are built in such a way that they are shaded during the extreme heat of the day, but receive direct sunlight in the cooler hours when the chicks can receive "solar brooding" through their dark skins; the skin of newly hatched young is blackish and absorbs solar heat, much as happens in the adults when they expose their dark skin and downy regions to sunlight to conserve energy (Ohmart 1973). Favoured nesting plants are chollas, especially the Staghorn Cholla *Opuntia versicolor* which also provides a densely branched support for the nest and spines to deter predators. The male broods the chicks at night and sleeps on the nest. He is relieved by the female shortly after daybreak and both parents then take turns to shade the young as T_A rises, oriented with the tail into the breeze, thereby channelling air down and over the chicks. Once the nestlings are 4-5 days old, they are able to thermoregulate to some degree and by 6 days the parents are absent from the nest for long periods during the heat of the day (07:00-13:00 h) when whiptailed lizards (*Cnemidophorus* spp.) are most active. These lizards are the food item most frequently fed to the young. The parents also regurgitate some food material at each feed, up to the time when the nestlings are about 14 days old, but the composition of this regurgitated material could not be analyzed. It looked like a clear, slightly viscous fluid and was delivered to the chick before the food item held in the parent's bill, the process taking from 25-195 s. Small, lethargic nestlings may be eaten by the parents if they do not beg when pecked at.

Hatching in the Roadrunner is asynchronous because incubation begins with the laying of the first egg of the clutch, as happens also in the Sociable Weaver of the Kalahari (Maclean 1973d). This distributes the food demands of the young and provides a mechanism whereby the youngest siblings in a nest succumb as conditions deteriorate, thereby reducing the brood size, as in the Sociable Weaver (Maclean 1973d). Brood reduction also occurs in the Curve-billed Thrasher *Toxostoma curvirostre* through starvation (Ricklefs 1965). The Lesser Roadrunner *G. velox* is a more tropical species which does not appear to have asynchronous hatching. A paucity of diurnal prey species has forced the Roadrunner to specialize in lizards which are large and brought to the nest one at a time. It is also important therefore that both parents are freed to hunt whiptailed lizards during the time of day when they are available, and for the nestling period to be prolonged in order to allow for fluctuations in food supply. The long nestling period is possible because the predation rate is low as a result of the spiny nest sites favoured by Roadrunners. The ability of adult Roadrunners to become hypothermic in winter is a further energy-saving measure when food is scarce, and their ability to absorb heat from the sun through specialized dark areas of skin and plumage augments this.

8.5 Larks

The nests of seven breeding species of larks in the Kalahari sandveld, South Africa, were oriented preferentially to lie on the eastern (44%), southeastern (20%) or southern (17%) side of the sheltering plant or other object; only 18% faced other sectors of the compass (Fig. 8.9; Maclean 1970a). Nests in exposed sites occurred only during a winter breeding season; summer nests were never exposed, nor were there any nests in the northwest (hottest) sector of the compass in summer. This was also the least used position in winter.

Most species laid a clutch of two or three eggs ($\bar{x}$ for 236 clutches = 2.17 eggs), but the proportion of three-egg clutches increased with increasing rainfall, especially among the nomadic seedeating species (Grey-backed Finchlark *Eremopterix verticalis*, Black-eared Finchlark *E. australis* (Fig. 7.5) and Pink-billed Lark *Spizocorys conirostris*). The overall breeding success (percentage of chicks which left the nest of eggs laid) was 17.9% over three breeding seasons. Once hatched, however, the young stood a better chance of survival than did the eggs: 46% of hatched young left the nest successfully. The main cause of mortality was predation by small mammals, snakes and small birds of prey.

Breeding of larks in the Namib Desert, Namibia, was confined almost entirely to the period March-May which coincided with the late-summer, early-autumn rains (Willoughby 1971). However, only Stark's Lark *Eremalauda starki* (Fig. 5.3) and the Grey-backed Finchlark appeared to have a definite gonadal cycle which corresponded with this wet season. The other four species of lark (Sabota Lark *Mirafra sabota*, Dune Lark

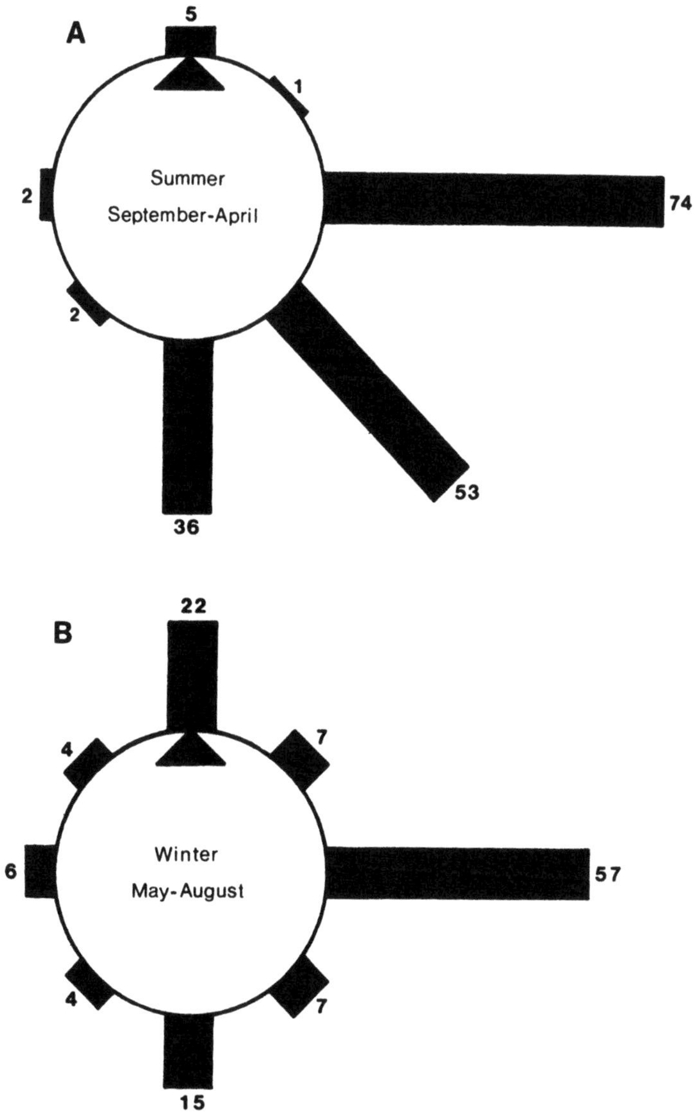

Fig. 8.9. Compass diagrams of orientation of nests of Kalahari larks, *A* in summer, *B* in winter, relative to sheltering object (usually a low plant). (After Maclean 1976)

M. erythrochlamys, Gray's Lark *Ammomanes grayi*, Spike-heeled Lark *Chersomanes albofasciata*) appeared to be able to breed at any time of the year, according to environmental conditions. The two gonadally seasonally cyclic larks also had cyclic moult patterns, which the other species seemed to lack, although somewhat meagre data show that the Sabota Lark may have a complete post-breeding moult between March and September. Of 31 nests of all species of larks in the Namib 22 nests (71%) were placed on the southern side of an object, eight nests (26%) on the southeastern side and only one nest faced east (Fig. 8.10). Similar figures were obtained for another small ground-nesting passerine in the Kalahari, namely the Lark-like Bunting *Emberiza impetuani*: of 32 nests, 24 (75%) faced south, seven nests (22%) faced southeast and only four nests (13%) faced east (Fig. 8.11; Maclean 1976c).

By contrast, in a study of nest temperatures in four nests of the Desert Lark *Ammomanes deserti* in the Negev Desert of Israel (Orr 1970), all the nests faced north, as would be expected in the northern hemisphere. Measurements were made after the nests had been abandoned. The temperature of the floor of the nests never exceeded 40.2 °C, which was 0.6 °C below the maximum ground temperature, but well above the maximum T_A of 34.8 °C recorded during the test period. During the day, the predominantly northwesterly wind, which was strongest during the hottest part of the day, helped further to keep nest temperatures moderate.

Incubation and nestling periods in the Kalahari larks were remarkably short, relative to those of arboreal birds of comparable body size: incubation took 12 days (minimum 11 days, maximum 13 days) and nestling about 10 days (Maclean 1970d). The duration of the nestling period was difficult to determine accurately, since young might leave the nest as early as 7 days old, especially if disturbed. The maximum nestling period for any of the seven species under investigation was 14 days for the Fawn-coloured Lark *Mirafra africanoides*; this species builds a domed nest, which may mean that the chicks are better protected from predation and therefore less pressured to leave the nest early.

During a study of the Dune Lark lasting 21 months (April 1985-December 1986), rainfall totalled 3.6 mm and fell only from early March to late June in the second year of study (Boyer 1988). Breeding may be initiated by the flush of insects which occurs at the time of the onset of egg-laying in this species. In 13 clutches, egg-laying occurred mostly in January and February, but some birds laid as early as August and as late as April. Nine (64%) of the 14 nests (one was not laid in) faced southeast. Completed clutches were of two eggs, incubation was by the female only and eggs were unattended while she was away foraging. Losses of eggs and nestlings were attributable to the causes shown in Table 8.2.

Of 18 eggs laid, 11 (61%) hatched, but 62% were lost at either the egg or nestling stage (Table 8.2), mainly as a result of predation. Chicks were fed mainly, if not entirely, on insects. Of the 11 hatched nestlings, only five survived to leave the nest (three from one pair and two from another), a success rate of 28% (young reared of eggs laid).

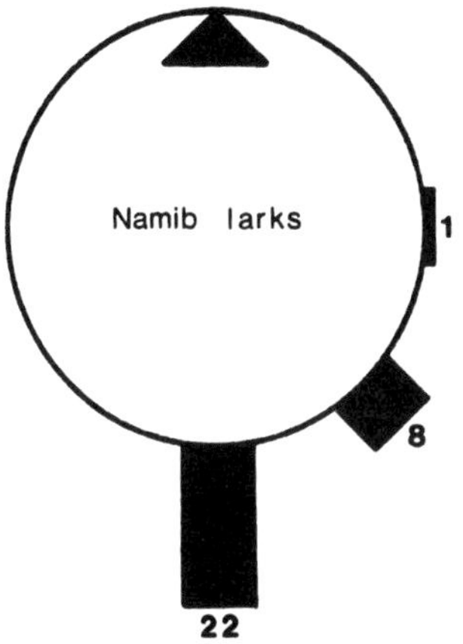

Fig. 8.10. Compass diagram of orientation of nests of Namib larks, relative to sheltering object (usually a low plant). (After Maclean 1976c)

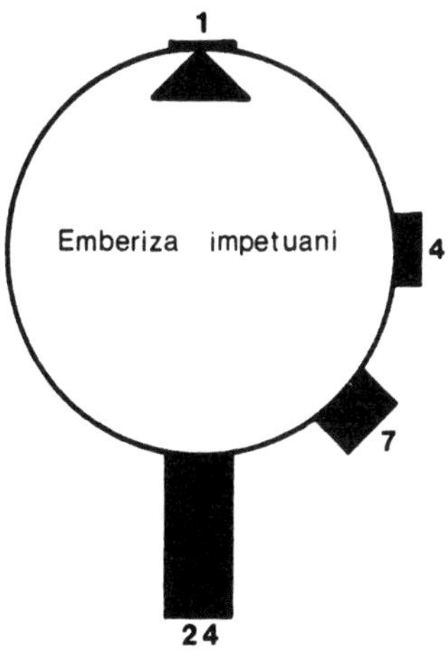

Fig. 8.11. Compass diagram of orientation of nests of Lark-like Bunting *Emberiza impetuani* in the Kalahari Desert relative to sheltering object (usuallly a stone). After Maclean 1976c)

Table 8.2. Causes of loss of eggs and nestlings in the Dune Lark. (Boyer 1988)

Cause of loss	Eggs	Nestlings	Eggs + nestlings	Loss (%)
Predation	3	3	6	46
Overheating	3	0	3	23
Infertility	1	-	1	8
Starvation	-	2	2	15
Unknown	0	1	1	8
Totals	7	6	13	100

The Dune Lark is the only passerine adapted to exist permanently in the dune sea of the Namib Desert, a hyperarid region with low, unpredictable rainfall. The birds do not drink, but their biology is otherwise poorly known. Longevity is a maximum of about 6 years, based on limited present knowledge; if each adult produced on average 0.5 young/year (Boyer 1988), a 5-year-old bird would theoretically produce 2.5 young, but productivity is apparently much lower than this: in the 1990 breeding season, only one of 13 pairs produced any young (Williams 1992). Nestlings are heavily preyed on by Black-backed Jackals *Canis mesomelas.*

8.6 The Sociable Weaver *Philetairus socius* (Fig. 3.5)

In a 19-month study of the Sociable Weaver, Maclean (1973d) found several breeding adaptations to the arid environment. As was shown in Chapter 7, rainfall (or some closely related phenomenon, such as food supply) initiated breeding activity, but the amount of rain governed the number of pairs which bred, the clutch size and the breeding success. The clutch size varied from 2-6 eggs, with a mean of 3.54 eggs out of 623 observed complete clutches. The largest monthly mean clutch size was 4.10 eggs at the height of a good wet season, while the lowest was 2.88 eggs near the end of a prolonged, 9-month breeding season. This was at a time when environmental conditions were deteriorating as a result of increasing T_A, accompanied by greater evaporation of soil moisture, desiccation of the vegetation and a drop in the supply of insect food. However, such inevitable events are partly set off by two factors:

1. Incubation in the Sociable Weaver starts with the first or second egg of the clutch, leading to asynchronous hatching of the chicks (Fig. 8.12); older chicks have the advantage of greater strength over the younger and, when food is in short supply, the younger members of a brood may die of starvation, quickly reducing brood size and enhancing the continued survival of the stronger chicks.
2. The young of earlier broods in a given season help their parents to feed the young of later broods, so that the young of a fourth brood may be fed by as many as 11 birds (the parents and the nine young of three previous broods).

Maclean (1973d) states that, therefore, "the parents need to do much less work toward the end of a breeding period, by which time they are in worn plumage and possibly somewhat exhausted from their exertion after several months of breeding activity. It is in any case useful to have as many workers feeding chicks as possible at a time when the food supply is dwindling."

The greatest minimum age of a Sociable Weaver ringed in this study and retrapped in subsequent years was 3 years 3 months; the mean minimum age of 12 Sociable Weavers retrapped at the same time was 2 years 10 months. The species is therefore potentially long-lived for a small passerine (mean weight 27 g; Maclean 1993), which accords with the need for the adults to survive several drought years (as many as 6 years in succession), when insufficient rain may fall to allow the birds to breed successfully.

8.7 Other Small Passerines

Three species of ground-nesting birds in North American grasslands (shortgrass prairie in north-central Colorado), the Lark Bunting *Calamospiza melanocorys*, McCown's Longspur *Calcarius mccownii* and Horned Lark *Eremophila alpestris* placed their nests so as to experience a reduction in wind velocity by at least an order of magnitude, but vegetation did not appear to act as a windbreak (With and Webb 1993). Nests of Lark Buntings were placed beneath shrubs and bunch grasses on the leeward side; the vegetation did not function as a windbreak, since the more exposed nests of McCown's Longspurs and Horned Larks surprisingly were more sheltered from wind than those of Lark Buntings. Nor did spatial orientation of the nests of any species necessarily coincide with prevailing wind conditions. Nests of longspurs and larks were almost completely exposed to the sun during the middle of the day (09:00-15:00 h) The placing of nests of Lark

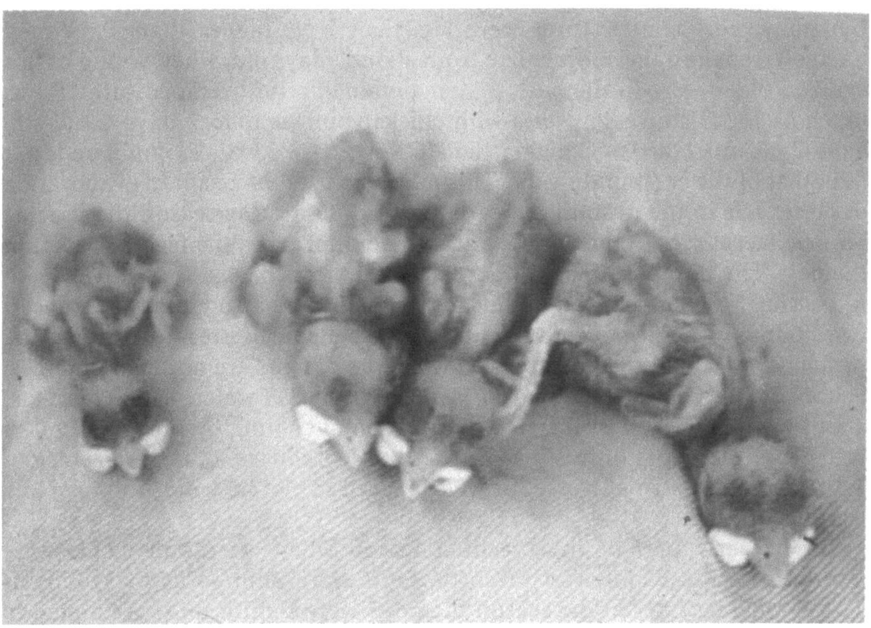

Fig. 8.12. Brood of young Sociable Weavers *Philetairus socius* showing discrepant ages resulting from incubation starting with the first egg of the clutch

Buntings was probably for greater shading (60% of the day as compared with only 45% of the day in the other two species). Similarly the nests of Lark Buntings were about 40% shaded during the midday hours as opposed to only 5% shaded for longspur nests and 13% shaded for lark nests during this period.

All three species had a non-random orientation of their nests: Lark Buntings and McCown's Longspurs had a significant northerly orientation, while Horned Larks had a significantly northeasterly orientation. The degree of radiative cover appears to be a more important consequence of nest placement than shelter from wind.

Lark Buntings began to breed later in the season than longspurs or larks and were therefore exposed to higher T_A and lower wind velocities. The Lark Bunting was unique among the three study species in that the males incubated; males of this species are black, so that nest placing may be important in providing opportunity for radiative cover and convective cooling. The earlier-nesting longspurs and larks may have had thermal benefit from solar radiation in their more exposed nest sites.

The daily fluctuation in T_A in the Lower Sonoran Desert near Tucson, Arizona (Fig. 1.4), may exceed 20 °C (in summer from a low of 15 °C to a maximum of 40 °C; Ricklefs and Hainsworth 1969). The foraging of adult Cactus Wrens *Campylorhynchus brunneicapillus* is depressed at T_A >35 °C.

Temperature within the bulky domed nests varied from ambient to as much as 6 °C higher, depending on how the sun was striking the nest. Generally, nest temperatures were less than 2 °C higher than T_A. Young Cactus Wrens can thermoregulate from about 7 days old; when young of this age and older were in the nest, they maintained a nest temperature (T_N) as much as 13 °C above T_A. Nests with chicks younger than 7 days could still have T_N as much as 10 °C higher than T_A. Above T_A 33 °C, T_N could be lower than that of the surrounding air. This may be partly the result of evaporation of faeces left in the nest: in the early part of the nesting season the parents remove faecal sacs of the young, but in the hotter parts the faeces are left in the nest. The number of faecal sacs left in the nest correlates positively with increasing T_A. Non-removal of the faecal sacs may be an adaptation to desert conditions, or it may be the result of physiological stress on the young which, when it is hot, defaecate even when the parents are not in attendance. Faeces of the young contained an average of 66.3% water; each chick lost about 1.4 ml of water every 12 h. Experimental cooling by pouring 3 ml of water into the floors of five recently used nests reduced T_N by an average of 1.3 °C, whereas a small vessel of water placed in the nest did not have any significant effect.

In the early part of the breeding season when afternoon winds were strong from the sector west-to-north, nests were built so as to avoid this sector, whereas later in the season the nests were significantly orientated to the southwest so as to face the wind during the hottest time of the day. Roosting nests were oriented randomly.

The most obvious advantage of the Cactus Wren's bulky closed nest is protection of the young from the sun. At higher T_A heat is effectively dissipated from the nest by convection or evaporation, while at lower T_A it is effectively retained. Orientation of the nest takes advantage of local climatic factors for the retention or loss of heat.

When Simons and Martin (1990) provided breeding Cactus Wrens with supplemental food in the form of mealworms during the nestling period, the wrens fed their nestlings more, increased the rate of food delivery and had greater nesting success than pairs without supplemental food. Of food-supplemented pairs, 86% had a second brood, as opposed to only 50% of pairs without food supplementation. The conclusion is that food supply limits reproductive rate and success in the Cactus Wren. Because of the relatively unpredictable nature of food supply in deserts, it is probable that brood size in the Cactus Wren has evolved to allow parents to take advantage of periods when food is more abundant. Food-supplementation experiments were not conducted at other stages of the life cycle, so it is not possible to say to what extent the food supply limits other aspects of reproduction.

The Rufous-eared Warbler *Malcorus pectoralis* (Fig. 7.8) is a typical small, arid-zone, insectivorous passerine of body mass 8-11.5 g ($\bar{x}$ of 10 birds = 10 g; Maclean 1993). It builds its nest in a variety of shrubs and bushes in the Kalahari, not selecting for thorny species: 43% of 28 nests were built in the thornless shrub *Rhigozum trichotomum*, 39% in other non-thorny plants and only 18% of nests in thorny species of *Acacia* (Maclean 1974a). As in the case of lark nests, the entrances of the nests of the Rufous-eared Warbler were oriented preferentially facing east (36%), south (28%) and southeast (16%). This was presumably so that the nest contents were maximally shaded during the hot hours of the day when the sun shines from the north, northwest and west.

Yom-Tov (1989) calculated that clutch size in endemic Australian passerine bird families (Monarchidae, Maluridae and Meliphagidae) is smaller on average (2.39 eggs) than that of "new invader" families (Ploceidae, Hirundinidae, *Corvus*; 4.40 eggs), regardless of the environments they inhabit. The explanation he offered is that either the endemics had low reproductive rates or that environmental factors were responsible. It is not clear what Yom-Tov really means here, but in any case his "phyletic" explanation that the endemics might have invaded Australia via the tropical forests to the north is doubtfully valid in the light of new evidence of the autochthony of these taxa. His "environmental" explanation suggests that erratic rainfall and drought, which are characteristic of much of Australia, could be controlling factors, but these do not explain the higher clutch sizes in the "new invaders" in the same habitats. He evades the issue by concluding that "there is more than one solution to a problem, and that animals living in the same environment may adopt different solutions to similar problems."

8.8 Conclusions

Several parental strategies enhance the survival of young birds in hot environments or in situations where wind may present problems. The orientation of the nest (or its entrance in the case of a closed nest) relative to a sheltering object provides shelter from sun and/or wind, depending on the greater need. Shielding the young and eggs from harsh environmental conditions is also performed by most parental birds, especially if overheating is probable in full sunlight. The body posture of birds nesting in open situations on the ground is often important in the regulation of egg temperature. Attentive periods in such species are usually short (1.5-2 h), but sandgrouse are unusual in that the female incubates throughout most of the daylight hours, the male during the night. The water-vapour conductance of eggs may be reduced, partly through reduced pore area in the shell. Provision of the young with water may occur by regurgitation, or in succulent food, or by soaked belly feathers (as in sandgrouse). Bringing water in the crop or plumage is a strategy open only to birds which fly frequently to drinking places, or which live near water. Asynchronous hatching of the young allows the brood size to be reduced quickly if food is suddenly in short supply. Similarly the evolution of helpers at the nest enhances the survival of nestlings as food supply dwindles towards the end of a breeding period.

Some birds may have a reduced clutch size to allow for breeding all through the year, without taxing the carrying capacity of the environment, even in the worst of times. In larks and perhaps other small birds, a small clutch size may also allow for rapid replacement of eggs in the face of high predation rates.

9 Into Tomorrow........

One of the most striking facts to emerge from this review of the literature on arid-zone ornithology is how the research has centred on the desert regions of North America, southern Africa, North Africa and Australia. The remaining deserts have received very little attention, yet they must have much to offer the researcher in the way of new and comparative information. Some work has been done in the deserts of South America and a little on those of Asia, both of which regions harbour arid-adapted birds of many kinds: the unique and poorly known seedsnipe (Thinocoridae) of South America and the ground jays *Podoces* of arid Asia, to name but two groups, yet there is still so much to do.

Time, however, is neither on the side of the birds nor on the side of the research workers. To quote Lovegrove (1993): "There can be no doubt that desert ecosystems are among the most threatened in the world and require special attention ....". The number of endemic birds in arid zones is proportionately very high, and the number of endemics in the ecotones between the deserts and their neighbouring biomes is possibly even higher, but the conserved areas do not by any means include all of these potentially vulnerable species. For example, Lovegrove (1993) lists no fewer than eight species of birds endemic to the northern arid zone of Namibia, which do not necessarily occur in the very large Etosha National Park: these species are Rüppell's Parrot *Poicephalus rueppellii*, Violet Woodhoopoe *Phoeniculus damarensis*, Monteiro's Hornbill *Tockus monteiri*, Rockrunner *Achaetops pycnopygius*, Carp's Black Tit *Parus carpi*, Bare-cheeked Babbler *Turdoides gymnogenys*, White-tailed Shrike *Lanioturdus torquatus* and Herero Chat *Namibornis herero*. To this list may also be added Hartlaub's Francolin *Francolinus hartlaubi*, which, along with the shrike and the chat, inhabits rocky hills and mountains, only some of which are in protected areas. The uniqueness of these nine endemics is highlighted by the fact that three of them belong to monotypic genera.

The level of endemism is significant also in the arid southwestern parts of Madagascar, a semidesert of Subarid Thorn Scrub (Langrand 1990) whose ornithological ecophysiology is all but unknown. Nicoll and Langrand (1989) remark that the avian species diversity in the Southern Domain of Madagascar is considerable, and the level of endemism is very high. This environment is, however, impoverished by fire and overgrazing, a fate all too common in arid regions inhabited by peasant cultures. Interestingly, though, of the 118 species of birds recorded in the Southern Domain of Madagascar, the eight endemics are all forest species, despite the arid nature of the habitat. Three are ground-dwellers (Subdesert Mesite *Monias benschi*, Running Coua *Coua cursor* and Long-tailed Groundroller *Uratelornis chimaera*) and the remaining five arboreal (Verreaux's Coua *C. verreauxi*, Littoral Rockthrush *Pseudocossyphus imerinus*, Thamnornis Warbler *Thamnornis chloropetoides*, Archbold's Newtonia *Newtonia archboldi* and Lafresnaye's Vanga *Xenopirostris xenopirostris*) (Langrand 1990). All these species are threatened by deforestation, while the ground-dwellers are further endangered by trapping, dogs and rats *Rattus rattus*.

In spite of these immediate threats to the wildlife of the arid parts of Madagascar, only a few reserves have been set aside for its conservation, but at least something is being done. In southern Africa, of whose area over half is classified as arid to semi-arid, the desert biome "enjoys an above-average conservation status" (Lovegrove 1993), which is encouraging. Lovegrove (1993) divides the arid parts of southern Africa into Desert, Nama Karoo and Arid Savanna; each of these biomes has three endemic species of birds. Desert is confined mainly to Namibia and harbours Rüppell's Korhaan *Eupodotis rueppellii*, the Dune Lark *Mirafra erythrochlamys* and Gray's Lark *Ammomanes grayi*. The Nama Karoo takes up much of the Cape Province (South Africa) and southern Namibia; its three endemics are the Red Lark *Mirafra burra*, Sclater's Lark *Spizocorys sclateri* and Cinnamon-breasted Warbler *Euryptila subcinnamomea*. The Arid Savanna includes most of the Kalahari sandveld of the Northern Cape Province, eastern Namibia and most of Botswana; endemic to this biome are the Violet Woodhoopoe, Pied Babbler *Turdoides bicolor* and Rockrunner.

The North American and Australian deserts also enjoy considerable conservation status, but this is certainly not true of the deserts of South America, North Africa and Asia. Not only are these latter regions being degraded by overexploitation by people, but their ecotonal borders are being encouraged to spread by the process called desertification. "Desertification is the degradation of lands in dry areas" (Grainger 1990). It leads to the spread of desert-like conditions into regions not formerly classifiable as desert, yet the process does not necessarily encourage the spread of desert organisms, since the very destruction of the habitat is inimical to the survival of most forms of wildlife. Desertification is not the spread of desert conditions, but rather an insidious process of aridification through poor land use, namely overcultivation, overgrazing, poor irrigation practices, and deforestation (Grainger 1990). The decline in biological productivity in desertified areas renders them unsuitable not only for wildlife, but for any form of agriculture or animal husbandry. The major impact of desertification on humankind has been felt in the Sahel of Africa and in western Rajasthan, India, where great loss of life, both human and livestock, has resulted in the past because of drought (Mainguet 1991). Exploitation of natural resources during good years means overexploitation in the bad years; the carrying capacity of deserts is limited to what they can support in the worst of times, and should not exceed that when conditions are at their best.

Yet in all the hundreds of thousands of words written about desertification, hardly any attention has been paid to the conservation of wildlife in arid regions. The causes of desertification are well known, and the remedies have been frequently aired, but nothing happens. Human populations continue to grow; the areas of mismanaged land grow even faster, and the productivity of the land diminishes apace. Aid organizations compete with each other for funds to send to the ravaged countries, but there is little or no cooperation or coordination between them and their activities, nor is there a concerted effort to educate the occupants of the degraded lands. Social customs, poor farming methods and political interests often conflict with conservation measures, so the deterioration goes on. This happens in first-world as well as in third-world countries: for example, the arid Karoo of South Africa has been subjected for many years to overexploitative farming methods, to the detriment of the soil, the vegetation and the birdlife. As a result, sensitive species, such as Burchell's Courser *Cursorius rufus* have shown a marked and accelerating decline in numbers in recent years.

Mainguet (1991), writing about the approach to the conservation of the desert environment, pointed out that the aim should be to rediscover the earth, its soil, its water and its vegetation; and also to remember that words are not actions and that diagnoses are not cures. If this rediscovery can be achieved, future research on birds of the arid zones may progress beyond the baselines established so far. More fieldwork and experimental work are needed to determine actual limits of tolerance to heat and water loss, sources of water, and water balance. We need to know how nomadic birds find their way to regions of recent rainfall, and what determines whether or not the birds will breed in a given place at a given time: is it the food supply governing body condition, or is it some indirect effect such as reproductive hormones in the food itself? What levels of such effects are needed to ensure the onset of breeding, as well as its continuation within a given season? What causes breeding to cease? The field of arid-zone ornithology is still wide open for the adventurous investigator.

References

Aldrich HC (1943) Some notes on the Common Sandgrouse (*Pterocles exustus* Temminck) in Kaira District. J Bombay Nat Hist Soc 44:123-125

Alkon PU, Degen AA, Pinshow B, Shaw PJ (1985) Phenology, diet and water turnover rates of Negev Desert Chukars. J Arid Environ 9:51-61

Amanova MB (1984) The ecological-physiological adaptations of water metabolism in desert birds. J Arid Environ 7:199-200

Arad Z, Gavrieli-Levin I, Eylath U, Marder J (1987) Effect of dehydration on cutaneous water evaporation in heat-exposed pigeons (*Columba livia*). Physiol Zool 60:623-630

Arad Z, Gavrieli-Levin I, Marder J (1988) Adaptation of the pigeon egg to incubation in dry hot environments. Physiol Zool 61:293-300

Bartholomew GA (1972) The water economy of seed-eating birds that survive without drinking. Proc 15th Int Orn Congr, The Hague, pp 237-254

Bartholomew GA, Cade TJ (1963) The water economy of land birds. Auk 80:504-539

Bartholomew GA, Dawson WR (1954) Body temperature and water requirements in the Mourning Dove, *Zenaidura macroura marginella*. Ecology 35:181-187

Bartholomew GA, Dawson WR (1979) Thermoregulatory behavior during incubation in Heermann's Gulls. Physiol Zool 52:422-437

Bartholomew GA, Hudson JW, Howell TR (1962) Body temperature, oxygen consumption, evaporative water loss, and heart rate in the Poor-will. Condor 64:117-125

Bartholomew GA, White FN, Howell TR (1976) The thermal significance of the nest of the Sociable Weaver *Philetairus socius*: summer observations. Ibis 118:402-410

Begg GW, Maclean GL (1976) Belly-soaking in the Whitecrowned Plover. Ostrich 46:65

Bennett PM, Harvey PH (1987) Active and resting metabolism in birds: allometry, phylogeny and ecology. J Zool (Lond) 213:327-363

Bernstein MH (1971) Cutaneous water loss in small birds. Condor 73:468-469

Boyer HJ (1988) Breeding biology of the Dune Lark. Ostrich 59:30-37

Cade TJ (1964) Water and salt balance in granivorous birds. In: Wayner MJ (ed) Thirst. Proc 1st Int Symp on Thirst in the regulation of body water. Pergamon Press, Oxford, pp 237-256

Cade TJ (1965) Relations between raptors and columbiform birds at a desert water hole. Wilson Bull 77:340-345

Cade TJ, Dybas JA Jr (1962) Water economy of the Budgerygah. Auk 79:345-364

Cade TJ, Greenwald LI (1966a) Drinking behavior of mousebirds in the Namib Desert, southern Africa. Auk 83:126-128

Cade TJ, Greenwald L (1966b) Nasal salt secretion in falconiform birds. Condor 68:338-350

Cade TJ, Maclean GL (1967) Transport of water by adult sandgrouse to their young. Condor 69:323-343

Cade TJ, Tobin CA, Gold A (1965) Water economy and metabolism of two estrildine finches. Physiol Zool 38:9-33

Cade TJ, Willoughby EJ, Maclean GL (1966) Drinking behavior of sandgrouse in the Namib and Kalahari Desert, Africa. Auk 83:124-126

Calder WA (1964) Gaseous metabolism and water relations of the Zebra Finch, *Taeniopygia castanotis*. Physiol Zool 37:400-413

Calder WA (1968) The diurnal activity of the Roadrunner, *Geococcyx californianus*. Condor 70:84-85

Casselton PJ (1984) Breeding birds. In: Cloudsley-Thompson JL (ed) Key environments: Sahara Desert. Pergamon Press, Oxford, pp 229-240

Chandola A, Bhatt D, Pathak VK (1983) Environmental manipulation of seasonal reproduction in Spotted Munia *Lonchura punctulata*. In: Mikami S, Homma K, Wada M (eds) Avian endocrinology: environmental and ecological perspectives. Japan Scientific Societies Press, Tokyo, pp 229-242

Chappell MA, Goldstein DL, Winkler DW (1984) Oxygen consumption, evaporative water loss, and temperature regulation of California Gull chicks (*Larus californicus*) in a desert rookery. Physiol Zool 57:204-214

Christensen GC (1963) The Imperial Sand Grouse in the Thar Desert of India. Occas Pap Biol Soc Nevada 2:1-7

Christensen GC, Bohl WH, Bump G (1964) A study and review of the Common Indian Sandgrouse and the Imperial Sandgrouse. US Dep Interior, Washington, DC

Cloudsley-Thompson JL (1979) Adaptive functions of the colours of desert animals. J Arid Environ 2:95-104

Cloudsley-Thompson JL, Mohamed ERM (1967) Water economy of the Ostrich. Nature 216:1040

Cox GW (1983) Foraging behaviour of the Dune Lark. Ostrich 54:113-120

Cramp S, Reynolds JF (1972) Studies of less familiar birds, 168: Cream-coloured Courser. Br Birds 65:120-124

Crawford EC, Lasiewski RC (1968) Oxygen consumption and respiratory evaporation of the Emu and Rhea. Condor 70:333-339

Davies SJJF (1976) Environmental variables and the biology of Australian arid-zone birds. Proc 16th Int Orn Congr, Canberra, 1974, pp 481-488

Davies SJJF (1977) The timing of breeding by the Zebra Finch *Taeniopygia castanotis* at Mileura, Western Australia. Ibis 119:369-372

Davies SJJF (1984) Nomadism as a response to desert conditions in Australia. J Arid Environ 7:183-196

Dawson WR (1954) Temperature regulation and water requirements of the Brown and Abert Towhees, *Pipilo fuscus* and *Pipilo aberti*. Univ Calif Publ Zool 59:81-124

Dawson WR (1976) Physiological and behavioural adjustments of birds to heat and aridity. Proc 16th Int Orn Congr, Canberra, 1974, pp 455-467

Dawson WR (1981) Adjustments of Australian birds to thermal conditions and water scarcity in arid zones. In: Keast A (ed) Ecological biogeography of Australia. Junk, The Hague, pp 1651-1674

Dawson WR (1984) Physiological studies of desert birds: present and future considerations. J Arid Environ 7:133-156

Dawson WR, Bartholomew GA (1968) Temperature regulation and water economy of desert birds. In: Brown GW (ed) Desert biology, vol 1. Academic Press, New York, pp 357-394

Dawson WR, Bennett AF (1973) Roles of metabolic level and temperature regulation in the adjustment of western Plumed Pigeons (*Lophophaps ferruginea*) to desert conditions. Comp Biochem Physiol 44A:249-266

Dawson WR, Fisher CD (1969) Responses to temperature by the Spotted Nightjar (*Eurostopodus guttatus*). Condor 71:49-53

Dawson WR, Hudson JW (1970) Birds. In: Whittow GC (ed) Comparative physiology of temperature regulation, vol 1. Academic Press, New York, pp 223-310

Dean WRJ, Milton SJ, Watkeys MK, Hockey PAR (1991) Distribution, habitat preference and conservation status of the Red Lark *Certhilauda burra* in Cape Province, South Africa. Biol Conserv 58:257-274

Dean WRJ, Williams JB, Milton SJ (1993) Breeding of the White-backed Mousebird *Colius colius* in relation to rainfall and the phenology of fruiting plants in the southern Karoo, South Africa. J Afr Zool 107:105-111

Dixon J, Louw G (1978) Seasonal effects on nutrition, reproduction and aspects of thermoregulation in the Namaqua Sandgrouse (*Pterocles namaqua*). Madoqua 11:19-29

Dixon JEW (1978) Animal remains recovered from sandgrouse (Aves, Pteroclidae) crops in the Etosha National Park. Madoqua 11:75-76
Drozdov NN (1984) The levels and aspects of convergent adaptations in desert birds. J Arid Environ 7:201-202
Evenari M (1985) The desert environment. In: Evenari M, Noy-Meir I, Goodall DW (eds) Ecosystems of the world, 12A. Hot deserts and shrublands. Elsevier, Amsterdam, pp 1-22
Fisher CD, Lindgren E, Dawson WR (1972) Drinking patterns and behavior of Australian desert birds in relation to their ecology and abundance. Condor 74:111-136
Frith HJ (1982) Pigeons and doves of Australia. Rigby, Adelaide
Frith HJ, Carpenter SM, Braithwaite LW (1976) Sexual cycles of pigeons in arid and semi-arid Australia. Aust J Zool 24:331-351
Frumkin R, Pinshow B, Weinstein Y (1986) Metabolic heat production and evaporative heat loss in desert phasianids: Chukar and Sand Partridge. Physiol Zool 59:592-605
Gatter W (1971) Wassertransport beim Flußregenpfeifer (*Charadrius dubius*). Vogelwelt 92:100-103
George U (1969) Über das Tränken der Jungen und andere Lebensäußerungen des Senegal-Flughuhns, *Pterocles senegallus*, in Marokko. J Ornithol 110:181-191
Ginn PJ (1977) Sandgrouse drinking behaviour in Botswana. Honeyguide 90:23-27
Goldstein DL, Williams JB, Braun EJ (1990) Osmoregulation in the field by salt-marsh Savannah Sparrows *Passerculus sandwichensis beldingi*. Physiol Zool 63:669-682
Grainger A (1990) The threatening desert: controlling desertification. Earthscan Publications, London
Grant GS (1982) Avian incubation: egg temperature, nest humidity, and behavioral thermoregulation in a hot environment. Ornithol Monogr 30:1-75
Gullion GW (1960) The ecology of Gambel's Quail in Nevada and the arid southwest. Ecology 41:518-536
Hamilton WJ, Heppner F (1967) Radiant solar energy and the function of black homeotherm pigmentation: an hypothesis. Science 155:196-197
Harvey PH, Pagel MD (1991) The comparative method in evolutionary biology. Oxford University Press, Oxford
Hinsley SA, Ferns PN (1994) Time and energy budgets of breeding males and females in sandgrouse *Pterocles* species. Ibis 136:261-270
Holmes WN, Phillips JG (1985) The avian salt gland. Biol Rev 60:213-256
Howell TR, Bartholomew GA (1959) Further experiments on torpidity in the Poor-will. Condor 61:180-185
Howell TR, Bartholomew GA (1961) Temperature regulation in Laysan and Black-footed Albatrosses. Condor 63:185-197
Howell TR, Bartholomew GA (1962) Temperature regulation in the Red-tailed Tropic Bird and the Red-footed Booby. Condor 64:6-18
Immelmann K (1963a) Drought adaptations in Australian desert birds. Proc 13th Int Orn Congr, Ithaca, New York, pp 649-657
Immelmann K (1963b) Tierische Jahresperiodik in ökologischer Sicht. Zool Jb (Syst) 91:91-200
Immelmann K (1967) Untersuchungen zur endogenen und exogenen Steuerung der Jahresperiodik afrikanischer Vögel. Verh Dtsch Zool Ges Heidelberg, Zool Anz Suppl 31:340-357
Immelmann K (1970) Environmental factors controlling reproduction in African and Australian birds - a comparison. Ostrich Suppl 8:193-204
Immelmann K (1971) Ecological aspects of periodic reproduction. In: Farner DS, King JR (eds) Avian biology, vol 1. Academic Press, New York, pp 341-389
Immelmann K (1972) Erörterungen zur Definition und Anwendbarkeit der Begriffe „Ultimate Factor", „Proximate Factor" und „Zeitgeber". Oecologia (Berl) 9:259-264
Inoue T (1963) Nasal salt gland: independence of salt and water transport. Science 142:1299-1300
Jaeger EC (1949) Further observations on the hibernations of the Poor-will. Condor 51:105-109
Jesson RA, Maclean GL (1976) Salt glands in the neonatal Australian Pratincole. Emu 76:277
Jones PJ, Ward P (1976) The level of reserve protein as the proximate factor controlling the timing of breeding and clutch-size in the Red-billed Quelea *Quelea quelea*. Ibis 118:547-574

Joubert CSW, Maclean GL (1973) The structure of the water-holding feathers of the Namaqua Sandgrouse. Zool Afr 8:141-152

Kalchreuter H (1979) Zur Mauser der äquatorialen Flughühner *Pterocles exustus* und *P. decoratus*. Bonn Zool Beitr 30:102-116

Keast A (1959) Australian birds: their zoogeography and adaptations to an arid continent. In: Keast A, Crocker RL, Christian CS (eds) Biogeography and ecology in Australia. Junk, The Hague, pp 89-114

Keast A (1968) Moult in birds of the Australian dry country relative to rainfall and breeding. J Zool (Lond) 155:185-200

Kemp AC (1973) Environmental factors affecting the onset of breeding in some southern African hornbills, *Tockus* spp. J Reprod Fertil Suppl 19:319-331

Kemp AC, Kemp MI (1972) A study of the biology of Monteiro's Hornbill. Ann Transvaal Mus 27:255-268

Kendeigh SC (1939) The relation of metabolism to the development of temperature regulation in birds. J Exp Zool 82:419-438

Kendeigh SC (1944) Effect of air temperature on the rate of energy metabolism in the English Sparrow. J Exp Zool 96:1-16

Kerley GIH (1991) Seed removal by rodents, birds and ants in the semi-arid Karoo, South Africa. J Arid Environ 20:63-69

Kleinhaus S, Pinshow B, Bernstein MH, Degen AA (1985) Brain temperature in heat-stressed, water-deprived desert phasianids:Sand Partridge (*Ammoperdix heyi*) and Chukar (*Alectoris chukar sinaica*). Physiol Zool 58:105-116

Langrand O (1990) Guide to the birds of Madagascar. Yale University Press, New Haven

Lasiewski RC, Bartholomew GA (1966) Evaporative cooling in the Poor-will and Tawny Frogmouth. Condor 68:253-262

Lasiewski RC, Dawson WR (1964) Physiological responses to temperature in the Common Nighthawk. Condor 66:477-490

Lasiewski RC, Bernstein MH, Ohmart RD (1971) Cutaneous water loss in the Roadrunner and Poor-will. Condor 73:470-472

Lee P, Schmidt-Nielsen K (1971) Respiratory and cutaneous evaporation in the Zebra Finch: effect on water balance. Am J Physiol 220:1598-1605

Louw GN, Belonje PC, Coetzee HJ (1969) Renal function, respiration, heart rate and thermoregulation in the Ostrich (*Struthio camelus*). Sci Pap Namib Des Res Stn IV. 1969 (No 42):43-54

Lovegrove B (1993) The living deserts of southern Africa. Fernwood Press, Vlaeberg (South Africa)

Lovegrove BG, Heldmaier G (1994) The amplitude of circadian body temperature rhythms in three rodents (*Aethomys namaquensis*, *Thallomys paedulcus* and *Cryptomys damarensis*) along an arboreal-subterranean gradient. Aust J Zool 42:65-78

Maclean GL (1967a) Die systematische Stellung der Flughühner (Pteroclididae). J Ornithol 108:203-217

Maclean GL (1967b) The breeding biology and behaviour of the Double-banded Courser *Rhinoptilus africanus* (Temminck). Ibis 109:556-569

Maclean GL (1968) Field studies on the sandgrouse of the Kalahari Desert. Living Bird 7:209-235

Maclean GL (1969) A study of seedsnipe in southern South America. Living Bird 8:33-80

Maclean GL (1970a) The biology of the larks (Alaudidae) of the Kalahari sandveld. Zool Afr 5:7-39

Maclean GL (1970b) The breeding seasons of birds in the south-western Kalahari. Ostrich Suppl 8:179-192

Maclean GL (1970c) An analysis of the avifauna of the southern Kalahari Gemsbok National Park. Zool Afr 5:249-273

Maclean GL (1970d) Breeding behaviour of larks in the Kalahari sandveld. Ann Natal Mus 20:381-401

Maclean GL (1973a) A review of the biology of the Australian desert waders, *Stiltia* and *Peltohyas*. Emu 73:61-70

Maclean GL (1973b) The Sociable Weaver, part 1. Description, distribution, dispersion and populations. Ostrich 44:176-190
Maclean GL (1973c) The Sociable Weaver, part 2. Nest architecture and social organization. Ostrich 44:191-218
Maclean GL (1973d) The Sociable Weaver, part 3. Breeding biology and moult. Ostrich 44:219-240
Maclean GL (1973e) The Sociable Weaver, part 4. Predators, parasites and symbionts. Ostrich 44:241-253
Maclean GL (1973f) The Sociable Weaver, part 5. Food, feeding and general behaviour. Ostrich 44:254-261
Maclean GL (1974a) The breeding biology of the Rufouseared Warbler and its bearing on the genus *Prinia*. Ostrich 45:9-14
Maclean GL (1974b) Arid-zone adaptations in southern African birds. Cimbebasia, Ser A 2:163-175
Maclean GL (1974c) Egg-covering in the Charadrii. Ostrich 45:167-174
Maclean GL (1975) Belly-soaking in the Charadriiformes. J Bombay Nat Hist Soc 72:74-82
Maclean GL (1976a) A field study of the Australian Pratincole. Emu 76:171-182
Maclean GL (1976b) A field study of the Australian Dotterel. Emu 76:207-215
Maclean GL (1976c) Arid-zone ornithology in Africa and South America. Proc 16th Int Orn Congress, Canberra, 1974, pp 468-480
Maclean GL (1976d) Adaptations of sandgrouse for life in arid lands. Proc 16th Int Orn Congr, Canberra, 1974, pp 502-516
Maclean GL (1983) Water transport by sandgrouse. BioScience 33:365-369
Maclean GL (1984a) Arid-zone adaptations of waders (Aves: Charadrii). S Afr J Zool 19:78-81
Maclean GL (1984b) Evolutionary trends in the sandgrouse (Pteroclidae). Malimbus 6:75-78
Maclean GL (1984c) Avian adaptations to the Kalahari environment: a typical continental semidesert. Koedoe Suppl 1984:187-193
Maclean GL (1985) Sandgrouse: models of adaptive compromise. S Afr J Wildl Res 15:1-6
Maclean GL (1993) Roberts' birds of southern Africa. John Voelcker Bird Book Fund, Cape Town
MacMillen RE, Trost CH (1966) Water economy and salt balance in White-winged and Inca Doves. Auk 83:441-456
Mainguet M (1991) Desertification: natural background and human mismanagement. Springer, Berlin Heidelberg New York
Marchant S (1958) The birds of the Santa Elena Peninsula, S.W. Ecuador. Ibis 100:349-387
Marchant S (1959) The breeding season in S.W.Ecuador. Ibis 101:137-152
Marchant S (1961) Observations on the breeding of the sandgrouse *Pterocles alchata* and *senegallus*. Bull Br Ornithol Club 81:134-141
Marchant S (1962) Watering of young in *Pterocles alchata*. Bull Br Ornithol Club 82:123-124
Marchant S (1963) The breeding of some Iraqi birds. Ibis 105:516-557
Marder J, Gavrieli-Levin I (1986) Body and egg temperature regulation in incubating pigeons exposed to heat stress: the role of skin evaporation. Physiol Zool 59:532-538
Marder J, Gavrieli-Levin I, Raber P (1986) Cutaneous evaporation in heat stressed Spotted Sandgrouse. Condor 88:99-100
Mares MA, Rosenzweig ML (1978) Granivory in North and South American deserts: rodents, birds and ants. Ecology 59:235-241
Mares MA, Blair WF, Enders F, Greegor D, Hulse A, Otte D, Sage R, Tomoff C (1977) Strategies and community patterns of desert animals. In: Solbrig O, Orians G (eds) Evolution in warm deserts. Dowden Hutchinson & Ross, Stroudsburg, pp 107-163
Marr TG, Raitt RJ (1983) Annual variations in patterns of reproduction of the Cactus Wren (*Campylorhynchus brunneicapillus*). Southwest Nat 28:149-156
Marschall U, Prinzinger R (1991) Vergleichende Ökophysiologie von fünf Prachtfinkenarten (Estrildidae). J Ornithol 132:319-323
Meade-Waldo EGB (1896) Sand grouse breeding in captivity. Zoologist 1896:298-299
Meade-Waldo EGB (1897) Sand grouse. Avic Mag 3 (New Ser):177-180
Milton S, Dean R, Siegfried R (1993) Ostriches eat anything: fact or fiction? Afr Wildl 47:205-208

Moldenhauer RR, Wiens JA (1970) The water economy of the Sage Sparrow, *Amphispiza belli nevadensis*. Condor 72:265-275
Morel G, Morel M-Y (1973) Recherches écologiques sur une savane sahélienne du Sénégal septentrional. Oiseaux: cycle annuel, biomasse et consommation. Ann Univ Abidjan, Sér E (Ecol) 6:147-155
Morel M-Y (1973) Cycles annuels de quelques oiseaux granivores des savanes africaines semi-arides. Bull IFAN 35:180-185
Morton SR, Davies PH (1983) Food of the Zebra Finch (*Poephila guttata*), and an examination of granivory in birds of the Australian arid zone. Aust J Ecol 8:235-243
Morton SR, MacMillen RE (1982) Seeds as sources of preformed water for desert-dwelling granivores. J Arid Environ 5:61-67
Nicoll ME, Langrand O (1989) Madagascar: revue de la conservation et des aires protégées. WWF, Gland (Switzerland)
Niethammer G (1959) Die Rolle der Auslese durch Feinde bei Wüstenvögeln. Bonn Zool Beitr 10:179-197
Nix HA (1976) Environmental control of breeding, post-breeding dispersal and migration of birds in the Australian region. Proc 16th Int Orn Congr, Canberra, 1974, pp 272-305
Ohmart RD (1973) Observations on the breeding adaptations of the Roadrunner. Condor 75:140-149
Ohmart RD, Chapman TE, McFarland LZ (1970) Water turnover in Roadrunners under different environmental conditions. Auk 87:787-793
Ohmart RD, Lasiewski RC (1971) Roadrunners: energy conservation by hypothermia and absorption of sunlight. Science 172:67-69
Orr Y (1970) Temperature measurements of the nest of the Desert Lark (*Ammomanes deserti deserti*). Condor 72:476-478
Parmenter RR, MacMahon JA, Vander Wall SB (1984) The measurement of granivory by desert rodents, birds and ants: a comparison of an energetics approach and a seed-dish technique. J Arid Environ 7:75-92
Peaker M, Linzell JL (1975) Salt glands in birds and reptiles. Cambridge University Press, Cambridge
Pianka ER, Huey RB (1971) Bird species density in the Kalahari and the Australian deserts. Koedoe 14:123-130
Pianka HD, Pianka ER (1970) Bird censuses from desert localities in Western Australia. Emu 70:17-22
Pinshow B, Degen AA, Alkon PU (1983) Water intake, existence energy, and responses to water deprivation in the Sand Partridge *Ammoperdix heyi* and the Chukar *Alectoris chukar*: two phasianids of the Negev Desert. Physiol Zool 56:281-289
Pohl-Apel G, Sossinka R (1975) Gonadenentwicklung beim Wellensittich *Melopsittacus undulatus* unter verschiedenen Lichtbedingungen. J Ornithol 116:207-212
Polis GA (1991) Desert communities: an overview of patterns and processes. In: Polis GA (ed) The ecology of desert communities. University of Arizona Press, Tucson, pp 1-26
Poulson TL, Bartholomew GA (1962) Salt balance in the Savannah Sparrow. Physiol Zool 35:109-119
Prinzinger R, Pressmar A, Schleucher E (1991) Body temperature in birds. Comp Biochem Physiol 99A:499-506
Purdue JR (1976) Thermal environment of the nest and related parental behavior in Snowy Plovers, *Charadrius alexandrinus*. Condor 78:180-185
Purdue JR, Haines H (1977) Salt water tolerance and water turnover in the Snowy Plover. Auk 94:248-255
Rautenberg W, May R, Necker R, Rosner G (1978) Control of panting by thermosensitive spinal neurons in birds. In: Piiper J (ed) Respiratory function in birds, adult and embryonic. Springer, Berlin Heidelberg New York, pp 204-210
Reichman OJ (1984) Spatial and temporal variation of seed distribution in Sonoran Desert soils. J Biogeogr 11:1-11
Ricklefs RE (1965) Brood reduction in the Curve-billed Thrasher. Condor 67:505-510
Ricklefs RE, Hainsworth FR (1968) Temperature dependent behavior of the Cactus Wren. Ecology 49:227-233

Ricklefs RE, Hainsworth FR (1969) Temperature regulation in nestling Cactus Wrens: the nest environment. Condor 71:32-37

Rounsevell D (1970) Salt excretion in the Australian Pipit *Anthus novaeseelandiae* (Aves: Motacillidae). Aust J Zool 18:373-377

Sauer EGF (1971) Zur Biologie der wilden Strauße Südwestafrikas. Z Köln Zoo 14:43-64

Sauer EGF, Sauer EM (1967) Verhaltensforschung an wilden Straußen in Südwestafrika. Umsch Wiss Tech 20:652-657

Schleucher E (1993) Life in extreme dryness and heat: a telemetric study of the behaviour of the Diamond Dove *Geopelia cuneata* in its natural habitat. Emu 93:251-258

Schleucher E, Prinzinger R, Withers PC (1991) Life in extreme environments: investigations on the ecophysiology of a desert birds, the Australian Diamond Dove (*Geopelia cuneata* Latham). Oecologia 88:72-76

Schmidt-Nielsen K (1964) Desert animals: physiological problems of heat and water. Clarendon Press, Oxford

Schmidt-Nielsen K, Borut A, Lee P, Crawford E Jr (1963) Nasal salt gland: independence of salt and water transport. Science 142:1299-1301

Schmidt-Nielsen K, Kanwisher J, Lasiewski RC (1967) Temperature regulation and respiration in the Ostrich. Fed Proc 26:600

Schmidt-Nielsen K, Kanwisher J, Lasiewski RC, Cohn JE, Bretz WL (1969) Temperature regulation and respiration in the Ostrich. Condor 71:341-352

Seely MK, Louw GN (1980) First approximation of the effects of rainfall on the ecology and energetics of a Namib Desert dune ecosystem. J Arid Environ 3:25-54

Seibert HC (1949) Differences between migrant and non-migrant birds in food and water intake at various temperatures and photoperiods. Auk 66:128-153

Serventy DL (1971) Biology of desert birds. In: Farner DS, King JR (eds) Avian biology, vol 1. Academic Press, New York, pp 287-339

Shkedy Y, Safriel UN (1991) Fat reserves of an opportunist and of a specialist species in the Negev Desert. Auk 108:556-561

Sibley CG, Monroe BL Jr (1990) Distribution and taxonomy of birds of the world. Yale University Press, New Haven

Sibley CG, Corbin KW, Ahlquist JE (1968) The relationships of the seed-snipe (Thinocoridae) as indicated by their egg white proteins and hemoglobins. Bonn Zool Beitr 19:235-248

Simons LS, Martin TE (1990) Food limitation of avian reproduction: an experiment with the Cactus Wren. Ecology 71:869-876

Skadhauge E (1972) Salt and water excretion in xerophilic birds. Symp Zool Soc Lond 31:113-131

Skadhauge E (1974) Cloacal resorption of salt and water in the Galah (*Cacatua roseicapilla*). J Physiol (Lond) 240:763-773

Skadhauge E (1981) Osmoregulation in birds. Springer, Berlin Heidelberg New York

Smyth M, Bartholomew GA (1966a) The water economy of the Black-throated Sparrow and the Rock Wren. Condor 68:447-458

Smyth M, Bartholomew GA (1966b) Effects of water deprivation and sodium chloride on the blood and urine of the Mourning Dove. Auk 83:597-602

Smyth M, Coulombe HN (1971) Notes on the use of desert springs by birds in California. Condor 73:240-243

Sopyiev O (1984) On some adaptations of birds due to the anthropogenic effects on the Kara-Kum Desert. J Arid Environ 7:197-198

Sossinka R (1972) Langfristiges Durstvermögen wilder und domestizierter Zebrafinken (*Taeniopygia guttata castanotis* Gould). J Ornithol 113:418-426

Sossinka R (1974) Der Einfluß von Durstperioden auf die Schilddrüsen- und Gonadenaktivität und ihre Bedeutung für die Brutperiodik des Zebrafinken (*Taeniopygia castanotis* Gould). J Ornithol 115:128-141

Sossinka R (1974/75) Vorzeitige sexuelle Reifung beim Zebrafinken als Anpassung an Trockengebiete. J SWA Wiss Ges 29:95-103

Sossinka R (1975) Quantitative Untersuchungen zur sexuelle Reifung des Zebrafinken, *Taeniopygia castanotis* Gould. Verh Dtsch Zool Ges 1974:344-347

Staaland H (1967) Anatomical and physiological adaptations of the nasal glands in charadriiformes birds. Comp Biochem Physiol 23:933-944

Sturkie PD (1976) Kidneys, extrarenal salt excretion, and urine. In: Sturkie PD (ed) Avian physiology. Springer, Berlin Heidelberg New York, pp 263-285

Sumner EL Jr (1935) A life history study of the California Quail, with recommendations for conservation and management. Calif Fish Game 21:167-342

Takei Y, Kobayashi H (1980) Angiotensin and drinking behavior in birds. In: Tanabe Y, Tanaka K, Ookawa T (eds) Biological rhythms in birds. Japan Scientific Societies Press, Tokyo, pp 287-299

Takei Y, Okawara Y, Kobayashi H (1989) Control of drinking in birds. In: Hughes MR, Chadwick A (eds) Progress in avian osmoregulation. Leeds Philosophical & Literary Society, Leeds

Thomas DH (1984a) Sandgrouse as models of avian adaptations to deserts. S Afr J Zool 19:113-120

Thomas DH (1984b) Adaptations of birds to desert conditions. Papers presented at a symposium during the 18th Int Ornithol Congr in Moscow on 23 Aug 1982. Introduction. J Arid Environ 7:131-132

Thomas DH (1984c) Adaptations of desert birds: sandgrouse (Pteroclididae) as highly successful inhabitants of Afro-Asian arid lands. J Arid Environ 7:157-181

Thomas DH, Maclean GL (1981) Comparison of physiological and behavioural thermoregulation and osmoregulation in two sympatric sandgrouse species. J Arid Environ 4:335-358

Thomas DH, Phillips JG (1978) The anatomy and physiology of the avian nasal salt glands. Pavo 16:89-104

Thomas DH, Robin AP (1977) Comparative studies of thermoregulatory and osmoregulatory behaviour and physiology of five species of sandgrouse (Aves: Pterocliidae) in Morocco. J Zool (Lond) 183:229-249

Thomas DH, Robin AP (1983) Description of the downy young of Lichtenstein's Sandgrouse *Pterocles lichtensteini* and the significance of "unpatterned" downy young in the Pteroclididae. Bull Br Ornithol Club 103:40-43

Thomas DH, Maclean GL, Clinning CF (1981) Daily patterns of behaviour compared between two sand-grouse species (Aves: Pteroclididae) in captivity. Madoqua 12:187-198

Thomas DH, Pinshow B, Degen AA (1984) Renal and lower intestinal contributions to the water economy of desert-dwelling phasianid birds: comparison of free-living and captive Chukars and Sand Partridges. Physiol Zool 57:128-136

Tomoff CW (1974) Avian species diversity in desert scrub. Ecology 55:396-403

Vleck CM (1993) Hormones, reproduction and behaviour in birds of the Sonoran Desert. In: Sharp PJ (ed) Avian endocrinology. J Endocrinol, Bristol, pp 73-86

Von Frisch O (1968) Küken, die man "säugen muß". Vogel-Kosmos 5:364-368

Von Frisch O (1969) Zur Jugendentwicklung und Ethologie des Spießflughuhns (*Pterocles alchata*). Bonn Zool Beitr 20:130-144

Vorhies CT (1928) Do southwestern quail require water? Am Nat 62:446-452

Vorhies CT (1945) Water requirements of desert animals in the southwest. Univ Ariz Tech Bull 107:487-525

Walsberg GE, Campbell GS, King JR (1978) Animal coat color and radiative heat gain: a re-evaluation. J Comp Physiol 126:211-222

Ward P (1972) The functional significance of mass drinking flights by sandgrouse: Pteroclididae. Ibis 114:533-536

Weathers WW (1981) Physiological thermoregulation in heat-stressed birds: consequences of body size. Physiol Zool 54:345-361

Whittow GC (1976) Regulation of body temperature. In: Sturkie PD (ed) Avian physiology. Springer, Berlin Heidelberg New York, pp 146-173

Wiens JA (1985) Vertebrate responses to environmental patchiness in arid and semiarid ecosystems. In: Pickett STA, White PS (eds) The ecology of natural disturbance and patch dynamics. Academic Press, New York, pp 163-193

Wiens JA (1991) The ecology of desert birds. In: Polis GA (ed) The ecology of desert communities. University of Arizona Press, Tucson

Wiens JA, Johnston RF (1977) Adaptive correlates of granivory in birds. In: Pinowski J, Kendeigh SC (eds) Granivorous birds in ecosystems. Cambridge University Press, Cambridge, pp 301-340
Williams JB (1992) Longevity of Dune Larks in the Namib Desert. Safring News 21:17-18
Williams JB, du Plessis MA (1995) Field metabolism and water flux of Sociable Weavers in the Kalahari Desert. Ibis (in press)
Williams JB, Pacelli MM, Braun EJ (1991) The effect of water deprivation on renal function in conscious unrestrianed Gambel's Quail (*Callipepla gambelii*). Physiol Zool 64:1200-1216
Williams JB, Siegfried WR, Milton SJ, Adams NJ, Dean WRJ, du Plessis MA, Jackson S (1993) Field metabolism, water requirements, and foraging behavior of wild Ostriches in the Namib. Ecology 74:390-404
Willoughby EJ (1966) Water requirements of the Ground Dove. Condor 68:243-248
Willoughby EJ (1968) Water economy of the Stark's Lark and Grey-backed Finch-lark from the Namib Desert of South West Africa. Comp Biochem Physiol 27:723-745
Willoughby EJ (1969a) Desert coloration in birds of the central Namib Desert. Sci Pap Namib Des Res Stn IV. 1969 (No 44):59-68
Willoughby EJ (1969b) Evaporative water loss of a small xerophilous finch, *Lonchura malabarica*. Comp Biochem Physiol 28:655-664
Willoughby EJ (1971) Biology of larks (Aves:Alaudidae) in the central Namib Desert. Zool Afr 6:133-176
Willoughby EJ, Cade TJ (1967) Drinking habits of birds in the central Namib Desert of South West Africa. Sci Pap Namib Des Res Stn V 1967 (No 31):1-35
With KA, Webb DR (1993) Microclimate of ground nests: the relative importance of radiative cover and wind breaks for three grassland species. Condor 95:401-413
Withers PC, Williams JB (1990) Metabolic and respiratory physiology of an arid-adapted Australian bird, the Spinifex Pigeon. Condor 92:961-969
Withers PC, Siegfried WR, Louw GN (1981) Desert ostrich exhales unsaturated air. S Afr J Sci 77:569-570
Wyndham E (1980a) Diurnal cycle, behaviour and social organization of the Budgerigar *Melopsittacus undulatus*. Emu 80:25-33
Wyndham E (1980b) Environment and food of the Budgerigar *Melopsittacus undulatus*. Aust J Ecol 5:47-61
Wyndham E (1980c) Diurnal changes in crop contents and total body lipid of Budgerigars *Melopsittacus undulatus*. Ibis 122:229-234
Wyndham E (1980d) Total body lipids of the Budgerigar, *Melopsittacus undulatus* (Psittaciformes: Platycercidae) in inland mid-eastern Australia. Aust J Zool 28:239-247
Wyndham E (1983) Movements and breeding seasons of the Budgerigar. Emu 82:276-282
Yom-Tov Y (1989) Environmental predictability and life history traits in arid environments. J Arid Environ 17:193-197

Subject Index

This book is dedicated to

Dr. Tom J. Cade

who gave me the opportunity to study arid-zone birds
in four continents and
whose enthusiasm was an example to me always,

and to

Cherie Maclean

whose help and support for nearly three decades
can never be repaid.

MIX
Papier aus verantwortungsvollen Quellen
Paper from responsible sources
FSC® C105338

If you have any concerns about our products,
you can contact us on
ProductSafety@springernature.com

In case Publisher is established outside the EU,
the EU authorized representative is:
Springer Nature Customer Service Center GmbH
Europaplatz 3, 69115 Heidelberg, Germany

Printed by Libri Plureos GmbH
in Hamburg, Germany